EXPErIMENTAL ELEMENTAry orGanic chemistry

JAMES A. MOORE

University of Delaware

SAUNDERS GOLDEN SERIES

W. B. SAUNDERS COMPANY Philadelphia • London • Toronto

W. B. Saunders Company: West Washington Square
Philadelphia, Pa. 19105

12 Dyott Street
London, WC1A   1DB

833 Oxford Street
Toronto, Ontario M8Z 5T9, Canada

Listed here is the latest translated edition of this book together
with the language of the translation and the publisher.

French (1st Edition) — Les Editions HRW
Ltee., Montreal, Canada

Front cover illustration from Strietwiesser, Andrew, Jr., Owens, Peter H.:
Orbital and Electron Density Diagrams: An Application of Computer Graphics.
Copyright 1973, Macmillan Publishing Co., Inc.

Experimental Elementary Organic Chemistry   ISBN 0-7216-6531-4

Last digit is the print number: 9   8   7   6   5   4   3   2

preface

The objectives of this organic chemistry laboratory text parallel those of the companion book, *Elementary Organic Chemistry*. The experiments have been chosen primarily with the needs and interests of health science majors in mind. Nearly half of the chapters focus on the traditional laboratory methods of organic chemistry; the remainder illuminate the chemistry of some important reactions and groups of compounds. Several experiments, among them TLC identification of analgesics (Chapter 7), temperature sensing with liquid crystals (Chapter 18), cross dyeing (Chapter 19), and identification using ir and nmr spectra (Chapter 21), are new to a laboratory text at this level. Some familiar and popular experiments, such as the isolation of caffeine and eugenol, have been retained.

Two firmly held beliefs will be evident. First, if a point can be made by a test tube observation or by preparation on a 100 mg scale, nothing is gained by carrying out the experiment with 10 or 20 times this amount of material, and time and chemicals can certainly be wasted. Thus, in demonstrating column chromatography (Chapter 6), or dehydration of alcohols (Chapter 10) or isolation of the constituents of milk (Chapter 15), only as much material as needed to make the observations is used.

The second point concerns the rather extensive detail in the procedures throughout the book. These experiments are intended for students who have had only a minimum of prior experience, and who have very little feel for the correct way to carry out laboratory manipulations. Laboratory time is brief and valuable, and students are not well served by terse instructions which amount to an invitation to rediscover the wheel. By providing explicit instructions on the equipment and techniques to be used, the student has a better chance to enjoy a successful result and to learn how to handle laboratory operations.

The final chapter on unknowns is adapted from the "open end" approach used in *Experimental Methods in Organic Chemistry*, and I am greatly indebted to my colleague David Dalrymple for his approval to pattern this and several other experiments on those in our earlier laboratory text. The identification of unknowns by a combination of chemical and spectral data has proved to be manageable and highly successful with large groups of students in a one-semester course. Two sets of unknowns selected for this book, each containing 50 compounds, are available from Aldrich Chemical Company; ir and nmr spectra are provided, with unlimited copying privileges, for each unknown.

Thanks are due to the many students and graduate instructors who have carried out and improved the experiments, and to Miss Kate Moore, who pioneered several of the new procedures. Professor Roger Murray has provided valuable advice and help throughout the preparation of the book. The advice of Professor William Weaver on dyeing, and of Dr. T. W. Davison on liquid crystals is gratefully acknowledged. Again, thanks are given to Mrs. Mary Ann Gregson for preparation of the manuscript.

JAMES A. MOORE

contents

one
 introduction: laboratory equipment and operations................. 1

two
 physical properties of organic compounds........................... 7

three
 crystallization.. 21

four
 distillation.. 31

five
 extraction... 41

six
 chromatography... 51

seven
 analysis of drugs by thin layer chromatography.................... 61

eight
 steam distillation of clove oil.................................. 71

nine
 isolation of caffeine... 79

ten
 reactivity in substitution reactions............................ 89

eleven
 reactions of alcohols... 99

twelve
 aromatic substitution... 113

thirteen
 the grignard reaction... 123

fourteen
sulfanilamide **131**

fifteen
chemistry of milk **139**

sixteen
carbohydrates **149**

seventeen
proteins and amino acids **163**

eighteen
lipids **173**

nineteen
dyes and dyeing **185**

twenty
infrared and nmr spectra **199**

twenty-one
identification of unknowns **207**

introduction: laboratory equipment and operations

Laboratory work is an integral and essential part of any chemistry course. Chemistry is an experimental science — the compounds and reactions that are met in lecture and classroom work have been discovered by *experimental observations*. Organic compounds exist as gases, liquids or solids with characteristic odors and physical properties. They are synthesized, and distilled, crystallized and chromatographed, and then transformed by reactions into other compounds. The purpose of laboratory work is to provide an opportunity to observe the reality of compounds and reactions, and to learn something of the operations and techniques that are used in experimental organic chemistry and in other areas where organic compounds are encountered.

LABORATORY SAFETY

Most organic compounds are flammable, and some have irritating or toxic vapors; many organic reactions are potentially violent. Some general precautions are given in this section, and specific hazards are mentioned in later chapters. Accidents are minimized by good sense and neat, orderly working habits, but they can occur, and you should be familiar with the location of fire extinguishers and any other items of safety equipment, and how and when to use them.

EYE PROTECTION. EYE PROTECTION MUST BE WORN AT ALL TIMES IN THE LABORATORY. Light-weight plastic safety glasses are available at low cost. Prescription glasses are acceptable; side-shields can be attached for protection against splashes. Contact lenses provide no protection; safety glasses must be used.

PERSONAL SAFETY. Never taste or inhale fumes of any chemical. Avoid contact of chemicals with the skin; wash hands as soon as possible after making transfers or other manipulations.

When inserting glass tubing into a rubber stopper or rubber tubing, lubricate with a drop of glycerine and protect hands with a towel. Do not use thin-wall transfer pipets as connectors for rubber tubing or stoppers; they are fragile and very easily crushed.

FIRST AID. When any chemical is spilled or splashed on the skin, the first step is to flush liberally with water for several minutes. Use an eyewash fountain if face or eyes are affected, or a sink faucet if an eye wash is not close by. For acid burns, sodium bicarbonate can be applied for first aid. For minor heat burns, cold water will lessen the pain; salves are not recommended. Report any burn or cut to your instructor.

FIRE HAZARDS. Most laboratory fires result from the ignition of solvent vapors with a burner flame. This hazard must always be kept in mind and the following important precautions should always be observed:

Do not heat organic liquids over a flame without using a condenser. Make sure that all joints are securely fitted in any distillation set-up.

Before lighting a flame, check to make sure that volatile liquids are not being poured or evaporated by someone in your vicinity. Before pouring solvents, check for flames in your vicinity.

Always turn off a burner as soon as you are finished using it.

As a general practice, particularly when a burner is in use, avoid loose-fitting long sleeves and cuffs; long hair should be tied back during laboratory work.

DISPOSAL OF CHEMICALS. Water-immiscible organic solvents and other liquids should be discarded in a designated waste-solvent can. They should never be poured into a sink if it can be avoided; if only a sink is available, flush thoroughly with water.

Water-insoluble solids and glass should be disposed of in a non-metallic chemical waste jar if available; do not throw them in a sink or waste paper basket.

PREPARATION FOR THE LABORATORY

It is essential for successful laboratory work that you understand, before beginning an experiment, what you are going to do, and why and how you are going to do it. Study the assigned chapter in advance, and plan your operations.

For effective laboratory work, it is most important to develop good working habits, and to learn the proper equipment for a given purpose and how to use it. Maintain a well-organized locker or equipment drawer; keep your equipment clean and as conveniently located as possible. Make a practice of washing or rinsing glassware as soon as it has been emptied. Take enough time to clean up and store equipment properly at the end of the day. Please remember not to put general equipment into your personal locker.

EQUIPMENT AND OPERATIONS

GLASSWARE.   Some of the equipment items that are commonly used in organic laboratory work are identified in Figure 1.1 to facilitate checking-in your locker equipment.

18×150 mm   25×100 mm
test tubes

beaker   Erlenmeyer flask   graduated cylinder   glass funnel

glassware for measuring and handling liquids or solutions

round-bottom flask   distilling flask   condensers (lower packed for fractional distillation)   distillation head   vacuum take-off adapter

glassware for distillation

side arm filter flask   Büchner funnel   Hirsch funnel (porcelain or plastic)   separatory funnel

equipment for suction filtration of solids

FIGURE 1.1 Glassware.

Test tubes, Erlenmeyer flasks and round-bottom flasks are used for handling solutions or liquids, and for carrying out most reactions. Solvents can be evaporated in these vessels (under reduced pressure if necessary), and they can be stoppered to permit shaking or storage. Beakers are much less suitable for these purposes and are useful mainly for handling or weighing solids or as a bath for heating or cooling. Use of the specialized items in Figure 1.1 for distillation, extraction and suction filtration will be explained in subsequent chapters.

HEATING. The two sources of heat in most undergraduate labs are the Bunsen burner and the steam bath. A burner is a convenient and rapid source of heat; it is also the cause of most laboratory fires. When heating a flask over a burner, place a wire gauze under the flask to distribute the heat, and be ever-vigilant for organic vapors.

Electrically heated mantles or baths are very desirable alternatives to Bunsen burners and, if available, should always be used instead of a flame. Electric hotplates are also useful, but these frequently contain a hot, exposed element and should never be used when organic vapors are present; in particular, *it is extremely hazardous to evaporate solvents on a hot plate.*

When possible, the steam bath (Fig. 1.2) should be used for heating or evaporating any organic liquid. Connect the bath with the steam source

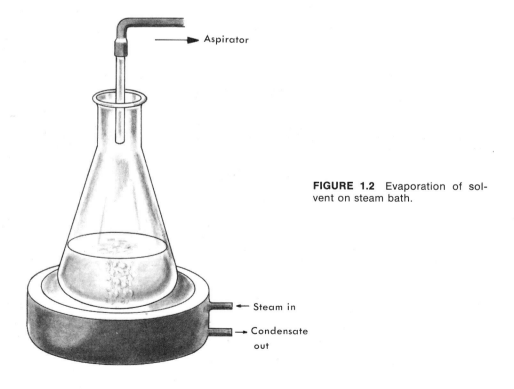

→ Aspirator

FIGURE 1.2 Evaporation of solvent on steam bath.

← Steam in

→ Condensate out

by means of the *upper* side arm; the tubing for release of condensate should be connected to the lower side arm. Place the flask to be heated on the largest ring that will support it. Do not remove the rings and set the flask on the bottom. Turn on steam until it just begins to escape; no benefit is gained from billowing clouds.

USE OF THE ASPIRATOR.   An aspirator (Fig. 1.3) provides a convenient source of reduced pressure for several operations. Water should be turned on to full capacity when using an aspirator. Splashing may be a problem; this can be avoided by tying a piece of rag around the outlet or wedging a piece of wire gauze in the trough below the stream. Heavy-wall rubber tubing should be used when connecting apparatus to the aspirator since regular tubing will collapse and pinch off the system. If the water pressure drops, or the water is turned off while the aspirator is in use, water tends to be drawn back through the arm into the evacuated system. To prevent this, a trap is connected between the aspirator and the evacuated vessel. Always release the vacuum *before* turning off the water.

HANDLING AND MEASURING CHEMICALS.   You will have many oc-casions to measure and transfer both solids and liquids. Occasionally a solid is free-flowing and can be poured from a bottle, but it usually should be removed with a spatula or scoop. A beaker is the most suitable vessel for weighing more than a few grams of solid. For weighing small quantities of reagents or products, glassine paper (not filter paper) should be used. The sample can then be placed in a vial or added to a solution simply by picking up the paper by opposite edges or corners and using it as an open funnel. Finely divided solids can be transferred quite completely by gently scraping with a spatula.

FIGURE 1.3 Aspirator and safety trap.

It is unnecessary to weigh reagents such as activated carbon or drying agents; the appropriate amount should be estimated by bulk. In this book, amounts of solids are sometimes given as volumes when accurate measurement is not required, in order to minimize time in weighing on a limited number of balances.

Liquid reagents and starting materials are usually measured in a graduated cylinder. In transferring liquids, do not attempt to pour by grasping a heavy bottle in one hand. For transferring a small volume of liquid, or filling a graduated cylinder to a specified mark, use a thin-wall transfer pipet and rubber bulb. These pipets are also useful for approximate measurements. Rubber bulbs of 1 ml and 2 ml capacity fill the pipet to about half and full capacity, respectively; the narrow section of these pipets contains about 0.2 ml. Approximate volumes of liquids can often be measured with adequate precision (± 10 to 15%) by the height of the liquid in a test tube. A guide for estimating volumes in test tubes is given on the inside back cover of this book.

PRODUCTS AND PERCENTAGE YIELD. In a number of experiments, products will be isolated from a natural source or from a reaction. In the former case, the percent yield is simply the weight of the product divided by the weight of the starting material $\times 100$. In a synthetic reaction, the percent yield is based on the theoretical amount of product that could be obtained from the starting compounds if the reaction were quantitative. In a reaction involving more than one starting compound, the theoretical yield is determined by the reactant which is used in smallest molar amount. In calculating the percentage yield, therefore, one must first determine the molar amounts of each starting compound and then the theoretical amount of product that could be obtained from the limiting reactant.

As an example, consider the esterification of succinic acid; 5.0 g of the acid is heated with 100 ml of ethanol, and 6.2 g of ester is isolated.

$$\text{HOCOCH}_2\text{CH}_2\text{CO}_2\text{H} + 2\ \text{C}_2\text{H}_5\text{OH} \xrightarrow{\text{H}^+} \text{C}_2\text{H}_5\text{OCOCH}_2\text{CH}_2\text{CO}_2\text{C}_2\text{H}_5 + 2\ \text{H}_2\text{O}$$

Succinic Acid	Ethanol	Diethyl Succinate

Calculating molar amounts:

Succinic Acid	Ethanol	Diethyl Succinate
5.0 g; mol wt = 118	100 ml; d = 0.79 g/ml	6.2 g; mol wt = 174
= 0.042 moles	= 79 g; mol wt = 46	= 0.036 moles
	= 1.72 moles	

The ethanol is present in large excess (0.084 moles required), and the yield of ester product is thus based on the acid. The theoretical amount of diethyl succinate is the same as that of acid, 0.042 moles, and the percentage yield is $\dfrac{.036}{.042} \times 100 = 86\%$.

physical properties
of organic compounds

One of the chief reasons that early chemists drew a distinction between organic and inorganic compounds was the difference in physical properties. Inorganic salts are generally high-melting solids, are soluble to some extent in water because of their ionic structure, but are insoluble in oils or other organic liquids. Many organic compounds are liquids; others are solids, usually insoluble in water, which melt or decompose at a fairly low temperature (30 to 200° C).

The **melting point** of a crystalline solid is the temperature at which solid and liquid exist in equilibrium. If heat is removed, some liquid solidifies, and if heat is added, some solid melts. Similarly, the **boiling point** of a compound is the temperature at which liquid and vapor exist in equilibrium. Energy is required to convert solid to liquid or liquid to gas. The changes in physical state on heating can be represented graphically, as shown in Figure 2.1.

MELTING POINTS OF SOLIDS

In Figure 2.1, the change from solid to liquid is shown as a sharp "break" or change in slope of the curve at the melting point. When this temperature is reached, heat is absorbed in breaking down the crystal lattice of the solid. If equilibrium between solid and liquid is maintained, the temperature does not rise until all of the solid is melted. In an actual melting point determination, the temperature increases somewhat in the time required for complete melting. The observed melting point is therefore usually a range of at least 1°.

A melting point can be easily and quickly determined on a very small sample, and it is a useful physical constant in characterizing a compound; i.e., it is one means of identification. However, the melting point is a single value, and hundreds of compounds can and do have the same melting point within a range of one degree. Absorption spectra (Chapter 20) provide much more information for characterizing and identifying compounds, but the melting point is still a useful place to begin.

7

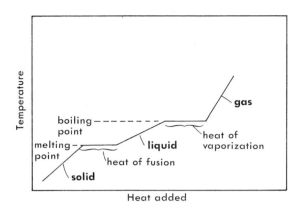

FIGURE 2.1 Heating curve.

Impurities in the crystal cause *broadening* and *lowering* of the melting point. Thus, a *mixture* of two compounds with the same melting point will usually have an unsharp melting range several degrees below the melting point of the pure compounds. This effect can be used as one indication of the identity of two substances that have the same melting point. If they are in fact the same compound, the melting point of an intimate mixture of the two will not be depressed. If a compound decomposes on heating, darkening and a broad melting range will be observed, since impurities are introduced by the decomposition.

A distinction must be made between *crystalline* and *amorphous* (noncrystalline) solids. An amorphous substance may be a hard, solid material; however, when it is heated only gradual softening occurs, with no distinct melting point. Glass is amorphous, and is actually a supercooled liquid which does not flow perceptibly, although ground glass may have a crystalline appearance.

Apparatus for Melting Point Determination. Determination of the melting point requires a thermometer in close contact with the sample and a means of heating the sample at a steady, controlled rate. This can be accomplished by holding a capillary tube containing the sample close to the thermometer bulb in a liquid bath (Fig. 2.2), or by placing the sample on a metal block in which the thermometer is imbedded (Fig. 2.3). The former method is preferable since each student can calibrate and use his own thermometer, and carry out the reading at his own desk. The melting point bath contains high-boiling oil, and can safely be heated to 250° with a burner. In a Thiele tube bath (Fig. 2.2a), convection currents keep the oil mixed when the bath is heated on the side loop.

Capillary tubes about 1.6 mm in diameter with one end sealed are usually purchased for use in melting point determinations. If the tubes have both ends open, heat the center lightly in a small flame until the glass collapses, and pull apart to form a pointed seal on each half; be sure that the end is completely sealed. Directions are given on page 13 for preparing very fine capillary tubes which are too small for use in melting points. Melting point capillaries can be prepared by pulling out a soft-glass test tube by the same procedure.

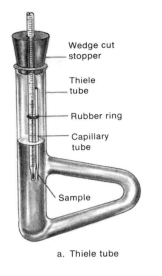

Wedge cut
stopper

Thiele
tube

Rubber ring

Capillary
tube

Sample

a. Thiele tube

FIGURE 2.2 Melting point baths.

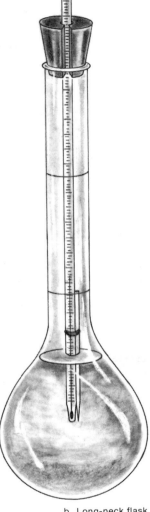

b. Long-neck flask

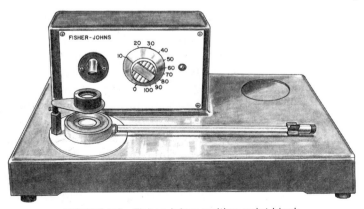

FIGURE 2.3 Fisher-Johns melting point block.

EXPERIMENTS

The objective of these experiments is to learn the technique of determining melting points, and to identify an unknown by melting point comparisons.

Melting Point of a Known Compound

Obtain a few crystals of ~~acetanilide~~ *naphthalene* on a watch glass or glassine paper. Pulverize the sample to a fine powder and scrape the material together into a little pile. Tamp the open end of a melting tube into the powder and then shake the solid down into the tube by tapping the sealed end on the desk top or gently stroking the tube with a file. A 2 to 3 mm (1/8″) column packed into the bottom of the tube is sufficient.

Attach the capillary tube to your thermometer with a small rubber band sliced from a piece of rubber tubing. Adjust the tube so that the end containing the sample is right next to the thermometer bulb, and place the thermometer in the bath at the position shown in Figure 2.2; be sure that the rubber ring is well above the level of the oil.

Heat the bath, fairly rapidly at first, until the temperature is about 15° below the expected melting point (113°). The heating rate must be reduced to about 2 to 3° per minute in the vicinity of the melting point, since heat transfer through the glass is poor. Keep your eye on the sample, with an occasional glance at the thermometer, and <u>record the temperature at which the sample begins to soften and become wet. Also record the temperature at which the last solid disappears.</u> Remove your thermometer, discard the capillary and allow the bath to cool.

Melting Point of an Unknown

Obtain a sample of an unknown assigned by the instructor; it will be one of the compounds listed in Table 2.1. Put a small amount of unknown aside, and then place some of the unknown into two capillary tubes. When the expected melting point is not known, it pays to heat rapidly (10 to 15° per minute) and observe an approximate melting range, and then repeat slowly with a second sample, starting about 15° below this temperature. Attach one of the tubes to the thermometer and determine the approximate melting point by rapid heating. Allow the bath to cool 15 to 20° while you attach the second sample, and then determine the melting point accurately. Record the melting point and, using Table 2.1, the names of the compounds that you believe your unknown could be, based on the melting point you observe.

TABLE 2.1

Compound	Melting Point	Compound	Melting Point
Benzhydrol	67°	p-Dibromobenzene	88°
Coumarin	69°	Benzil	94°
Biphenyl	70°	o-Anisic acid	100°
Phenylacetic acid	78°	o-Toluic acid	103°
Naphthalene	80°	m-Toluic acid	109°
Vanillin	81°	Resorcinol	110°
		Acetanilide	113°

Mixture Melting Point

After deciding which compound is most likely to be your unknown, obtain a sample and determine the melting point of a mixture of the compound and your unknown. Mix approximately equal amounts of the two samples together and powder them. Fill a tube with the mixture and another tube with the known compound. Attach them side-by-side to the thermometer so that the melting points of the pure compound and the mixture can be observed together. (A third tube containing unknown can also be included if desired.) Observe the melting point and decide whether your unknown and the known are the same. If not, try another candidate.

PHYSICAL PROPERTIES OF LIQUIDS

The boiling point of a liquid at a specified pressure is a characteristic physical constant, but since the boiling point varies with pressure, it is a less convenient and useful property than the melting point of a solid. Boiling points are usually determined by distillation of the liquid, as discussed in Chapter 4.

Liquids have several distinctive physical properties in addition to their change of state to gases on heating or to solids on cooling. These include density, refractive index, surface tension and viscosity. Organic liquids often differ considerably from water in these properties, and some of these differences are of practical importance in handling liquids.

The density of a liquid is the mass per unit volume, and is determined by weighing an accurately measured volume. The densities of organic compounds range from about 0.65 g/ml for alkanes to 3.3 g/ml for CH_2I_2. Liquids are commonly measured by volume, and the density must of course be known to determine the weight and molar amount. A point to bear in mind is that some organic solvents are less dense than water and when mixed with water form an *upper* layer; others are more dense and form the *lower* layer in a mixture with water.

The refractive index of a liquid is the ratio of the velocity of light in air to that in the liquid. The difference in velocity causes light rays to be refracted or change direction when they pass from air to liquid. When a rod

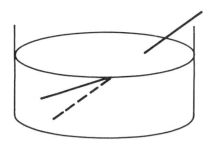

FIGURE 2.4 Refraction.

is partly submerged in water, refraction causes it to appear bent at the surface (Fig. 2–4). The swirling effect that is observed when two miscible liquids are poured together is caused by the changing refractive index as the liquids mix.

The refractive index of a liquid can be measured very accurately with an instrument called a refractometer, and the value is of some use as a physical constant. The refractive index of water is 1.33; values for organic compounds range from 1.3 to over 1.6 for some aromatic compounds. If the refractive index of a liquid is higher than about 1.5, the dispersion of light into a band of color can be seen when a flat-bottom glass bottle containing the liquid is held to the light at eye level. Diamonds have a very high refractive index (2.4), and it is this property that imparts "fire" to a faceted stone.

Surface tension is a force due to intermolecular attraction at the surface of a liquid; it is this force which holds liquid droplets in a spherical shape, causes liquids to "climb" by capillary attraction, and makes it possible for a needle to float on water. The surface tensions of liquids can be compared very simply by observing the height of liquid columns in capillary tubes. The surface tension of organic compounds is much lower than that of water, and organic liquids therefore form smaller drops and spread further than water on a solid surface. Addition of soap or other detergents to water lowers the surface tension, making it possible for a film of water to penetrate beneath an oily drop on a surface and dislodge it.

Viscosity is the resistance of a liquid to flow. It can be measured by the rate at which a drop of liquid passes through a capillary tube. Organic compounds vary tremendously in viscosity. Relative to water, with a value of about 10 on a scale of viscosity, alkanes and many other liquids have viscosities of 2 to 5. Alcohols have viscosities higher than water; glycols and glycerol have viscosities from 100 to over 1000.

EXPERIMENTS

The objective of these experiments is to gain experience in making observations by comparing the physical properties of a few typical organic compounds.

Obtain 1 to 2 ml samples of the following liquids in clean, dry 1 dram vials (label the vials first, and stand them in a small beaker): (1) acetone,

(2) hexane, (3) chlorobenzene and (4) 2-propanol. The structural formulas of the four compounds are:

smell color

O
‖
CH₃—C—CH₃

CH₃(CH₂)₄CH₃

Cl

OH
|
CH₃CHCH₃

Acetone   Hexane   Chlorobenzene   2-Propanol

Record your data in the report sheet as you make your observations and measurements.

Preliminary Observations

a. Using 10×75 mm test tubes, add a few drops of each liquid to about 1 ml of water and observe whether the liquid is less dense than water, more dense than water or dissolves in water. *floats*

Sinks b. Observe each of the liquids in the vial to see whether the refractive index is sufficiently high to cause visible dispersion.

Surface Tension

Preparation of Capillary Tube. A fine capillary tube is needed to compare surface tensions. This type of capillary will also be used later in thin layer chromatography, and is prepared by drawing out a section of the large end of a soft glass pipet (or a piece of 6 mm soft-glass tubing) that has been softened in a flame. Hold the pipet or tubing just above the inner blue cone of the flame, slowly rotating it back and forth. Continue heating, without pulling on the tubing, until the heated section is quite pliable and begins to sag or twist. Then remove from the flame and pull with a smooth motion until your hands are about 1½ to 2 feet apart. Select a 6 to 8 inch length from the center of the capillary section with the most uniform diameter and score the glass at each end of this section by gently stroking with a rough edge of clay plate or a piece of carborundum (boiling stone). Bend slightly at these marks so that the capillary breaks to give a smooth, squared-off end.

Measurement. Dip the end of the capillary tube just below the level of the acetone and hold it there for a few seconds. Withdraw the tube and measure the height of the liquid column on a metric scale to the nearest millimeter (see inside back cover of book). Record the value. Touch the tip to a piece of paper towel or filter paper to drain, and repeat the procedure several times. The column height should be within a 15 to 40 mm range. If it is less than 15 mm, a smaller bore capillary should be used; if it is more than 40 mm, a larger bore is needed.

Proceed in the same way with hexane and then with chlorobenzene, using the same end of the capillary each time. When you begin with a different compound, rinse the tube twice by dipping and draining before making the first measurement.

In the capillary tube, surface tension is the force that supports the column of liquid against the force of gravity; the liquid rises until the opposing forces are balanced. The force of gravity depends on the weight of the liquid in the column, or volume times density. Since the same tube is used for all measurements, the diameter is the same and the volume is proportional to the height. To compare surface tensions, the height of the column with each compound must be multiplied by the density. The densities of the compounds are given in the report sheet; multiply by the average height, and then compare the relative surface tensions with the values given.

Comparison with Water. To compare the surface tension of organic liquids and water, use a 1.6 mm melting point tube with both ends open (make a scratch and break off the sealed end if open tubes are not available). Measure the height of the liquid column using acetone, and then water. Calculate height times density and compare the surface tensions as in part 2a.

Effect of Detergent. Add a few grains of detergent or a few chips of soap to some water in a small beaker and shake to dissolve. Measure the height of the column of this solution in the melting point tube and compare with that of water.

Viscosity

Relative Viscosities of Acetone and 2-Propanol. The same capillary tube used for the surface tension measurement can be used. Dip the tube in acetone and then invert it and allow the liquid to run to the other end. Drain by touching to paper and repeat. Then dip the tube and adjust the height of the column to 1 cm by touching briefly to paper. Invert the tube partially, quickly stand it at an angle in a small beaker, full end up, and estimate the time required for the liquid to run to the lower end. Repeat several times, adding a little more liquid to keep a 1 cm column. If the tube is very fine, the fall can be timed with a sweep second hand.

Repeat the procedure with 2-propanol and compare the rate of fall with that of acetone. As in the case of the surface tension measurement, the viscosities of two liquids are proportional to the time of fall and the density of the compounds.

Viscosity of Glycerol. Obtain a drop of glycerol on a watch glass. Using the open end melting point capillary, compare the relative viscosities of acetone, water and glycerol.

crystallization

Organic compounds that are solids at room temperature are usually isolated and purified by **crystallization.** A compound is often obtained in crude form as a solution, from either a synthetic reaction or a natural source. The substance may be recovered from the solution simply by evaporating to a dry residue, but if other materials are present, there will be no purification or fractionation. In crystallization, however, the desired compound is allowed to separate by *selective crystal growth*, and impurities are retained in the mother liquor. The crystals are then collected by filtering the mixture. In recrystallization, a crude solid is dissolved in a suitable solvent, any insoluble impurities are filtered out, and then the compound is crystallized by cooling, or by evaporating some of the solvent.

Crystallization depends primarily on solubility relationships. With a few exceptions, the solubility of a compound in any solvent increases markedly with temperature, often manyfold over a range of 40 to 50°. This means that if a compound has adequate solubility (a few percent or more) in a hot solvent, it can often be crystallized quite completely just by cooling to a temperature at which the solution is supersaturated, *i.e.*, the solubility is exceeded.

The solubility of crystalline organic compounds depends on the functional groups that are present and the polarity of the solvent. "Like dissolves like," and compounds will be more soluble in solvents which are of a similar nature. Compounds with groups such as —OH, —NH—, —CONH and so forth are usually more soluble in solvents like alcohols (ROH) or water than in hydrocarbons. Regardless of the type of compound, however, the more stable the crystal lattice (the higher the melting point), the less soluble the compound, as seen in the melting points and solubilities of the isomeric nitrobenzoic acids.

	CO_2H with NO_2	CO_2H with NO_2	CO_2H with NO_2
	ortho	meta	para
Melting point	147°	141°	242°
Solubility, g/100 ml, in:			
Ethanol	28	33	2.2
Ether	21	25	0.9

GENERAL PROCEDURE FOR RECRYSTALLIZATION

1. The crude solid is dissolved in slightly more than the minimum amount of a suitable solvent, often with heating. Any impurities that remain undissolved, such as inorganic salts or bits of dirt, are removed by filtering the solution through paper or a small wad of cotton stuffed loosely in the bottom of a funnel. This step is frequently necessary to remove a drying agent that has been added to take up water. If the solution is hot, this filtration step must be carried out rapidly to avoid cooling and premature crystallization.

In addition to insoluble material, other contaminants can sometimes be removed from the solution in this clarification step by adding activated charcoal before filtration. Higher molecular-weight impurities, which often are responsible for dark color, are selectively *adsorbed* on the carbon particles. The decolorizing action of carbon will be seen in this experiment, but it is by no means always as effective as in this example. It is usually more successful in aqueous solutions than in organic solvents.

Filtration of a solution to remove impurities and/or carbon is best done with a fluted paper, which provides maximum surface. To prepare a fluted paper, lay a 4 inch filter paper disc on the benchtop, and fold it in half and then quarters. Reopen the latter fold like a book [Fig. 3.1(a)] and fold corners A and C up to meet point B. Reopen to a semicircle [Fig. 3.1(b)], and fold corner A first to point D and then to point E, and corner C to D and E, reopening to a semicircle after each fold. The paper should now look like Figure 3.1(c), with the folds bending the semicircle into a partially open cone. In each of the eight segments, make a fold in the center, alternating in the opposite direction to the previous folds, accordion fashion, to give a fan-like arrangement. When opened at the first fold (AC), a fluted paper [Fig. 3.1(d)] results.

2. The clarified solution is cooled, or, if necessary, concentrated by evaporation, or diluted with a miscible solvent in which the desired compound is less soluble. If crystals do not form spontaneously, scratching the wall of the flask just above the surface will often initiate crystallization.

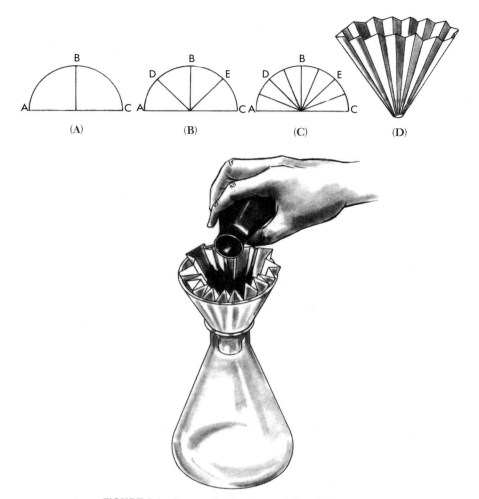

FIGURE 3.1 Preparation and use of fluted filter paper.

The size of the crystals depends on the rate of formation and the number of particles present; shaking the solution speeds up the process and gives smaller crystals.

3. The crystals are collected by swirling and pouring the mixture of solid and liquid into a suction filter (Fig. 3.2). The filtrate, or *mother liquor*, containing the soluble impurities (and of course some of the desired compound also) is drained off, and the crystals are then washed with a small amount of the solvent; this latter step is particularly important when the crystallization has occurred in a concentrated, viscous solution.

EXPERIMENTS

The objectives of these experiments are to observe crystallization and gain experience in purifying compounds by recrystallization.

FIGURE 3.2 Büchner funnel and filter flask.

Purification of Sugar

The purification of sucrose (table sugar) by crystallization is a major industry. Juice from cane or sugar beets is first treated with lime to remove proteins and acid impurities. The filtrate from this step is evaporated until crystallization of the crude sugar occurs; the dark mother liquor is processed as molasses. Further purification is accomplished by a recrystallization step which includes treatment with charcoal to remove color and strong flavor. In this experiment, brown sugar, which is impure sucrose, will be crystallized by adding ethanol, in which sucrose has a low solubility, to a concentrated aqueous solution. The resulting solution contains more dissolved sucrose than the equilibrium solubility limit, and the excess sugar then crystallizes.

Procedure. In a 13 × 100 mm test tube place enough brown sugar to form a roughly spherical mass (height equal to the diameter of the tube). Add 0.3 ml of water (10 drops from a transfer pipet). Heat a small beaker half full of water nearly to boiling, turn off the flame and place the test tube in the bath. Stir with a glass rod if necessary until all solid has dissolved.

Obtain 10 ml of ethanol, and with a pipet and bulb, add ethanol to the warm sugar syrup in 1 ml portions, shaking or stirring after each addition until the syrup and ethanol are thoroughly mixed. After adding 6 ml of ethanol, most of the colored impurities will be precipitated as a gum or sticky solid.

Place a very small tuft of cotton (about the size of a pea, *before* the cotton is compressed) in a glass funnel and pack it into the top of the stem.

Filter the ethanolic sugar solution through the cotton into a clean, dry test tube; rinse the emptied tube with the remainder of the alcohol and filter. If the filtrate still contains solid particles, press the cotton a little more firmly into the funnel and filter again. Dip the stirring rod into the solution, scrape it along the wall. Cork the tube and allow it to stand. Observe and record the appearance of the tube after an hour, and if possible, after it has stood for several days.

Preparation and Purification of Acetanilide

In this experiment, a product will be prepared and isolated by crystallization. The reaction is the acetylation of aniline, represented in the following equation. We will not be concerned with the nature or mechanism of the reaction at this point, but the yield of product can be calculated from the amounts of reactants and products.

Aniline	Acetic anhydride	Acetanilide	Acetic acid
d = 1.02 g/ml	d = 1.08 g/ml	mp = 114°	
bp = 184°	bp = 136°		
mol wt = 93	mol wt = 102	mol wt = 135	

Procedure. Measure 4 ml of aniline into a 10 ml graduated cylinder. Use the dropper attached to the bottle on the side shelf, and avoid contact of the aniline with the skin. Pour the aniline into a 250 ml Erlenmeyer flask. After draining the cylinder as completely as possible, fill it with water (10 to 12 ml) and pour this into the flask to obtain a few more droplets of oily aniline (some droplets will remain). Repeat this operation, adding a total of 25 ml of water to the Erlenmeyer flask containing the aniline.

In the same (wet) cylinder, measure 5 ml of acetic anhydride. (*Note:* Normally one would *not* use wet glassware to obtain a reagent; in this case it is permissible since water and aniline are present in the reaction.) Pour the acetic anhydride in several small portions into the mixture of aniline and water; swirl the contents briefly after each portion and record any changes that are observed. Allow the flask and contents to cool to room temperature (this can be speeded up by swirling in cold water).

The crude acetanilide is now recrystallized in the same flask. Add 100 ml of water and place the flask on a wire gauze which is securely supported on an iron ring; adjust the ring on the ring stand to allow about 1 inch of clearance above the top of a Bunsen burner. On a piece of paper, obtain about one-half teaspoonful of activated charcoal. Heat the flask contents to boiling with a burner flame; stir occasionally with a glass rod or spatula. When all of the oily material has dissolved (a trace of oil may remain on the wall above the solution), remove the flame, allow the solution to stop boiling, and add the charcoal. (If charcoal is added to the boiling solution, it will froth out of the flask.) Stir the hot mixture for several seconds and then resume heating with a low flame just sufficient to maintain minimal boiling. Also heat about 50 ml of water to boiling in a beaker for use in washing.

Place a fluted filter paper in a glass funnel and set it up for filtration into a 250 or 500 ml Erlenmeyer flask; pour a few milliliters of hot water through to warm the glass. Pick up the hot solution using a towel or other protection and pour it into the filter as rapidly as possible without overflowing the paper. If crystals begin to clog the filter or the stem of the funnel, stop pouring, add some boiling water to the filter to dissolve the crystals and reheat the solution of product to boiling before resuming the filtration. After the mixture has been filtered, rinse the flask with a little hot water and pour it through the filter. If a perceptible amount of carbon has leaked through the filter, giving a gray cast to the filtrate, reheat the solution to boiling and filter again through fresh paper. Allow the solution to cool and crystallize.

Prepare a suction filter using a porcelain Büchner funnel and a 250 ml heavy-wall side-arm filter flask (Fig. 3.2). Attach the arm to the aspirator side-arm with heavy-wall rubber tubing. Obtain a circle of filter paper just large enough to seal the crack between the edge of the disc and the glass funnel. The paper should not be so wide that it creases into a channel at the edge. Moisten the paper with water, turn on the aspirator and check that the paper is properly fitted.

Complete the crystallization by swirling the flask in an ice bath for a few minutes, and then collect the crystals on the suction filter. After completing the transfer, rinse the crystals with a little water and press the crystals with a spatula or glass stopper to remove as much water as possible. Before discarding the filtrate and washings, measure the volume with a graduated cylinder and record this in your report. Spread the crystals on paper to dry in your desk until the next laboratory period.

When the crystals are dry, record the weight, calculate the percentage yield and determine the melting point of your product.

distillation

Distillation is the process of vaporizing a liquid, condensing the vapor and collecting the condensate in a separate receiver. By this means, mixtures of compounds with different volatilities can be separated, or recovered from nonvolatile contaminants. When vaporization occurs from a solid, the process is called sublimation. The term evaporation is often used in laboratory practice when the residue, rather than the distillate, is of importance and the distilling vapors are permitted to escape uncondensed into a hood or an aspirator. The term reflux means vaporization with return of the condensate to the original flask.

For vaporization of a liquid to occur, the molecules must have sufficient kinetic energy to escape from the surface. The pressure exerted by the vaporization of a liquid in a closed container is the vapor pressure. Kinetic energy increases with temperature, and the vapor pressure therefore increases as a liquid is heated, as shown in Figure 4.1. When this pressure equals that of the atmosphere, boiling occurs, and the temperature at this pressure is the boiling point. Since the boiling point depends on the pressure, the temperature required for distillation can be reduced by lowering the pressure in the apparatus with an aspirator or vacuum pump.

In the distillation of a pure liquid, the boiling point remains constant as long as liquid and vapor are in equilibrium. With a *mixture* of two liquids A and B that have different vapor pressures, the vapor contains a higher proportion of the more volatile component, as seen in Figure 4.2. In this diagram, the boiling points of compounds A and B are T_A and T_B, and the boiling points of various mixtures of the two liquids are given by the lower curve. The composition of the vapor in equilibrium with any liquid is given by the point where a horizontal line, corresponding to the boiling point of the liquid, intersects the upper curve to the left.

When an equimolar mixture of A and B (composition X_1) is heated to the boiling point (T_1 on the lower curve), the vapor in equilibrium at this temperature is at point X_2 on the upper curve. Condensation of the vapor will give an initial distillate with composition X_2, richer in the more volatile compound A. If a sample of this vapor is removed, however, the liquid contains more of the higher boiling compound B. As distillation continues, the boiling point increases along the lower curve and the distillate contains progressively more B. A plot of the distillate volume versus temperature would resemble the dashed curve in Figure 4.3. In a simple distil-

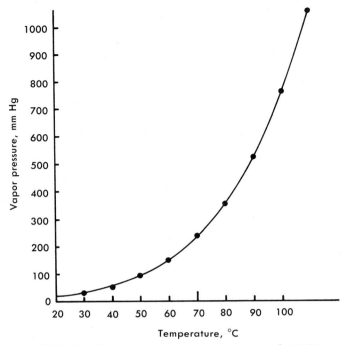

FIGURE 4.1 Vapor pressure–temperature curve for water.

lation of this type, there is relatively little separation of the two compounds unless the boiling points differ by 50° or more.

To improve the separation of A and B, the initial distillate, with composition X_2, can be condensed and redistilled. In this step the vapor in equilibrium with the liquid X_2 has the composition X_3, further enriched in

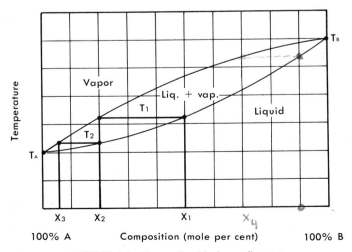

FIGURE 4.2 Vapor–liquid phase diagram.

$X_4 = 30\% A$
$70\% B$

FIGURE 4.3 Distillation curves.

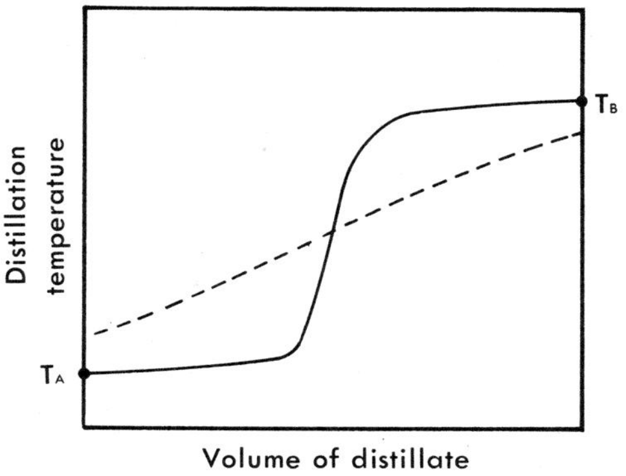

compound A. With a sufficient number of steps, essentially pure A can be obtained in the distillate. To accomplish repeated distillation steps in one operation, the distillation is carried out in a fractionating column, which contains a large surface area for constant equilibration between liquid and vapor. With such a column, packed with small turns of glass or coarse stainless steel wool, a distillation curve resembling the solid line in Figure 4.3 can be obtained.

APPARATUS FOR DISTILLATION

The basic elements of a distillation apparatus are a boiling flask, a column through which the vapor rises, a thermometer, a condenser and a receiver. Standard set-ups for distillation with and without a fractionating column are illustrated in Figures 4.4 and 4.5. These set-ups employ ground-glass-jointed equipment. If this is not available, the connections between flask, condenser and adapter can be made with smoothly bored, snug-fitting corks as shown in Figure 4.6.

In assembling ground-glass–joint apparatus, a very thin film of stop-cock grease is used to lubricate the joints; too much grease results in contamination of any liquid which contacts the joint, and the excess can prevent a tight seal. It is very important to be sure that the joint is clean, free of grit and properly aligned without stress. Always wipe the joint before assembling, and disassemble immediately after use.

To set up the apparatus, clamp the neck of the flask containing the liquid (with a boiling stone) to be distilled on a ring stand (Fig. 4.4). The height depends on what type of receiver is used and whether or not a fractionating column is included. Place a wire gauze on a ring, bring the ring up under the flask and tighten it on the stand so the flask is supported on the gauze. For simple distillation (Fig. 4.4), fit the distillation head into the neck of the flask and attach the rubber collar to hold the thermometer. For fractional distillation (Fig. 4.5), place the packed column in the boiling flask and clamp the column to the stand so that it is perfectly vertical. Fit the distillation head to the top of the packed column.

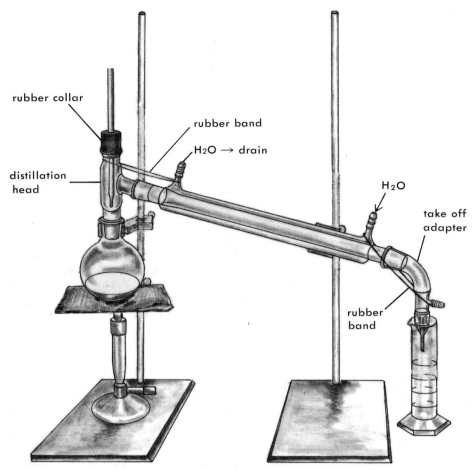

rubber collar

rubber band

H₂O → drain

distillation head

H₂O

take off adapter

rubber band

FIGURE 4.4 Apparatus for simple distillation.

Clamp the condenser to a second ring stand, supporting the weight at the midpoint with the fixed side of the clamp, and screwing down lightly with the movable side. Adjust the height and angle of the condenser to match that of the side-arm on the distillation head. Attach rubber tubing to the two side tubes on the condenser, using water to lubricate; connect the lower tube to a water tap and run the upper tubing out to the sink. Attach the take-off adapter to the lower joint of the condenser by means of a rubber band.

Bring the column assembly and condenser together and connect them, temporarily loosening the clamp around the condenser if necessary and adjusting its height so that the joint fits *without stress or binding*. A rubber band can be looped around the condenser side-arm and over the adapter to ensure a snug fit. Finally, loosen the clamp around the neck of the flask and raise the ring and wire gauze slightly so that the gauze exerts a slight pressure on the bottom of the flask to keep the joint tight and to prevent the escape of flammable vapors.

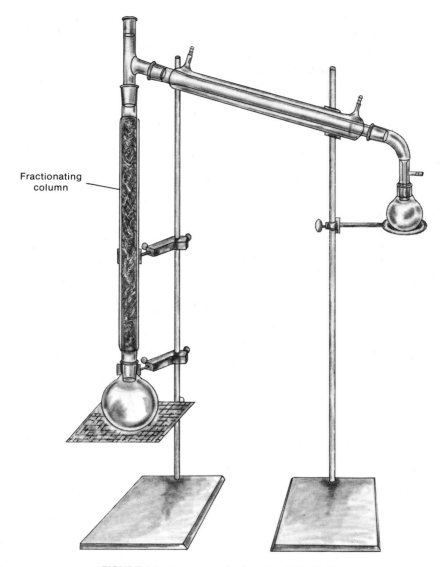

Fractionating
column

FIGURE 4.5 Apparatus for fractional distillation.

Place a suitable receiver (round-bottom flask, vial or graduated cylinder) under the adapter; if necessary, clamp it in position—do not prop the receiver on a make-shift support. Finally, insert a thermometer through the rubber collar with the bulb positioned just below the side-arm as shown in Figure 4.4.

A boiling stone is essential in any distillation to maintain constant ebullition (formation of bubbles and uniform turbulence). For this purpose a small lump of carborundum or other porous mineral is used. Without this stone to provide a steady stream of gas bubbles, superheating and bumping

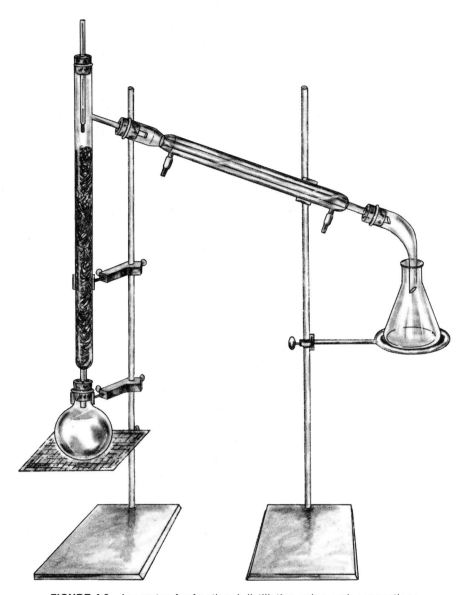

FIGURE 4.6 Apparatus for fractional distillation using cork connections.

of the liquid will usually occur. This applies with even greater force when evaporating a liquid on a steam bath without a condenser. If the boiling stone is forgotten until the liquid is hot, and possibly superheated, the flask must be cooled safely below the boiling point before adding the stone, or the liquid may erupt. Since the pores fill with liquid as soon as boiling ceases, the stone cannot be reused, and a fresh one must be added if the distillation is interrupted for any reason.

EXPERIMENTS

The objective of these experiments is to learn the procedure for distillation and to compare the results of simple and fractional distillation.

1. Simple Distillation

In a 100 ml round-bottom flask, place 40 ml of trichloroethylene (bp 87°) or another compound specified by the instructor. (Trichloroethylene is a common dry-cleaning solvent; it is nonflammable and relatively non-toxic in comparison to the extremely toxic carbon tetrachloride.) Assemble the apparatus as shown in Figure 4.4; use a 50 ml graduated cylinder as a receiver. Make sure that a boiling stone is present in the liquid and that all joints are secure. Turn on a gentle, steady flow of water through the condenser jacket.

Distillation Procedure. Position a burner under the gauze, light it and adjust to a low, steady flame. When the liquid has been heated to boiling, a ring of condensing vapor will be seen rising in the flask and distilling head. The thermometer will then register a rapid increase. Record the temperature when condensate begins to collect in the condenser. Adjust the rate of heating so that the thermometer is bathed in vapor, with a drop of condensate on the tip. Record the temperature when steady distillation is underway, and continue the distillation at a rate of 2 to 3 ml per minute.

Note: The temperature in the distilling head will not level off to the boiling point until thermal equilibrium with the glass walls has been established. This requires a minute or two, and at least a few drops of distillate will be collected before the boiling point is registered. If heating is too rapid, an appreciable amount of distillate may appear to be lower boiling "forerun." If distillation is too slow, particularly with a high boiling liquid, the true boiling point may never be observed because of heat losses.

Stop the distillation when the volume of liquid in the flask is about 2 to 3 ml (a puddle 1/2 inch in diameter). (Do not let the flask become dry, since without the absorption of heat by means of vaporization, the flask temperature can rapidly rise to several hundred degrees.) Pour the distillate and the residue into designated containers. Turn off the water and drain the condenser to avoid condensation of moisture.

2. Distillation of a Mixture

Obtain 40 ml of a mixture of two compounds designated by the instructor. Assemble the apparatus for simple distillation and follow the procedure given in Part 1. During this distillation, record the boiling point of the mixture at the beginning of distillation and then after each 5 ml portion of distillate is collected. You may find it necessary to increase the heating somewhat to maintain steady distillation. Save the distillate and residue for Part 3.

Plot a curve of distillation temperature *vs.* volume in the report sheet, and estimate, if possible, the boiling points of the two compounds in the mixture.

3. Fractional Distillation

Pour the distillate from Part 2 back into the same flask (with residue) and add a fresh boiling stone. Assemble the apparatus with a packed fractionating column as shown in Figure 4.5.

Repeat the distillation as outlined in Part 2. A somewhat larger flame may be needed to obtain steady distillation. The packed column should be wet with liquid but not flooded with condensate. Record the initial boiling point and the temperature after each 5 ml of distillate, as before. When the distillation is completed, place the distillate and residue in designated containers. After dismantling the apparatus, rinse the fractionating column with a few milliliters of acetone and then hold the aspirator tubing at one end and pull a stream of air through to dry it out.

On the same graph used for Part 2, plot the distillation curve and again estimate the boiling points of the compounds.

distribution coefficient

$K_D = C_1/C_2$

extraction

Extraction is the general term for recovery of a substance from a crude solid or a solution by bringing it into contact with a solvent that preferentially dissolves the desired material. In the isolation of organic compounds from a plant source, extraction of the dried leaf, bark or wood is commonly the first step. In synthetic organic chemistry, a reaction product is frequently obtained as a solution or a suspension in water, along with inorganic and organic by-products. By shaking the aqueous mixture with a water-immiscible organic solvent such as ether or chloroform, the product is transferred to the organic layer and can then be recovered by evaporation of the solvent.

The extraction of a compound from an aqueous solution using an organic solvent, or *vice versa*, is an equilibrium process governed by the solubilities of the substance in the two phases. The ratio of solubilities in the two solvents is called the **distribution coefficient, $K_D = C_1/C_2$,** which is an equilibrium constant with a characteristic value for any compound at a given temperature.

Let us consider the extraction of a compound whose solubilities in ether and water are 10 g/100 ml and 2 g/100 ml, respectively. If a solution of 1 g of the compound in 100 ml of water is extracted with 100 ml of ether, the fraction of the compound transferred to the ether phase can be calculated as follows:

$$K_{e/w} = \frac{C_{ether}}{C_{water}} = \frac{10/100}{2/100} = 5$$

Let x = g in ether at equilibrium

1 − x = g in water

$$\frac{x/100}{(1-x)/100} = 5$$

$$x = 5 - 5x$$

x = 0.83 g in ether

1 − x = 0.17 g in water

If the extraction is carried out with the same amount of ether in two equal portions, however, we have for the first extraction:

$$\frac{x/50}{(1-x)/100} = 5;$$

$x = 0.71$ g in ether

$1 - x = 0.29$ g in water

and for the second extraction (x' = g in ether at equilibrium):

$$\frac{x'/50}{(0.29 - x')/100} = 5;$$

$x' = 0.21$ g in ether

$x + x' = 0.92$

The total amount extracted by 100 ml of ether is thus 0.92 g.

It is seen from these values that virtually complete removal of the compound can be effected, even if the distribution coefficient is very low, by repeated extractions with small volumes of solvent. In practice this is accomplished by use of an apparatus in which the solution to be extracted is continuously treated with fresh solvent (Fig. 5.1). Most applications of liquid-liquid extraction in the laboratory require only a few contacts with fresh portions of solvent.

USE OF A SEPARATORY FUNNEL

The separatory funnel is a tapered vessel with a stopcock at the bottom which permits a sharp separation of two liquid layers in a liquid-liquid extraction or in any situation requiring the separation of an organic liquid from the aqueous layer. A separatory funnel is expensive and fragile, and when full, it is top heavy. The funnel should be supported on a ring of the proper size at a convenient height; don't prop it up on its stem. Before each use, check that the stopcock is seated and rotates freely. A clip or leash should be used to prevent the stopcock from falling out if it is accidentally loosened. A *very light* film of stopcock lubricant should be applied around the stopcock in bands on each side of the hole. (Teflon stopcocks require no lubricant.) Excess grease will be washed away by organic solvents and contaminate the solution.

The separatory funnel should be filled to no more than about three-fourths of the total depth, so that thorough mixing is possible. After filling (check first that stopcock is closed!), stopper securely with a properly fitting glass, plastic or rubber stopper. Before using the funnel for the first time, it is a good idea to shake with a few milliliters of solvent to make sure that stopcock and stopper are tight.

Hold the funnel with the stopcock end tilted up; the stopper is kept in place securely with the heel of one hand and the stopcock end is supported in the other hand (Fig. 5.2). As soon as the funnel is inverted, open the stopcock to release any pressure. Then close the stopcock and shake in

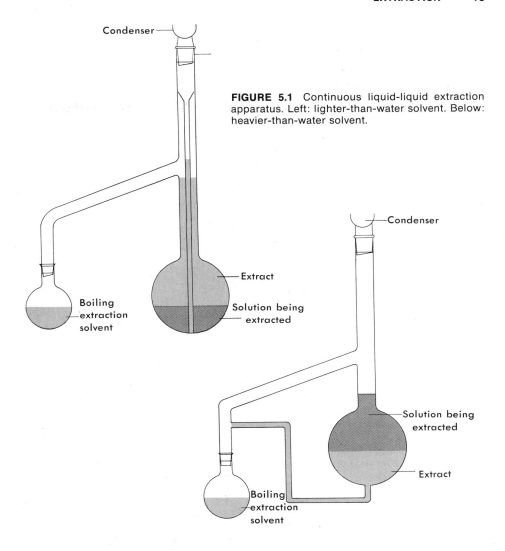

FIGURE 5.1 Continuous liquid-liquid extraction apparatus. Left: lighter-than-water solvent. Below: heavier-than-water solvent.

a horizontal position for about 1 minute. Stop, and slowly open the stop-cock a few times to vent any pressure. Replace the funnel in the ring and loosen the stopper. When the phases have completely separated, draw off the lower layer through the stopcock.

In the extraction of an aqueous solution, the solvent may be either lighter than water (e.g., ether or hexane) or heavier than water (e.g., chloroform or methylene chloride). In the first case, if several portions of solvent are used, the aqueous layer must be drained into a receiver (usually the flask in which it was originally contained), and the ether solution is transferred to a second flask. The aqueous phase is then returned to the separatory funnel for further extraction as needed. With a solvent denser than water, the aqueous solution is simply retained in the funnel and shaken with successive portions of solvent.

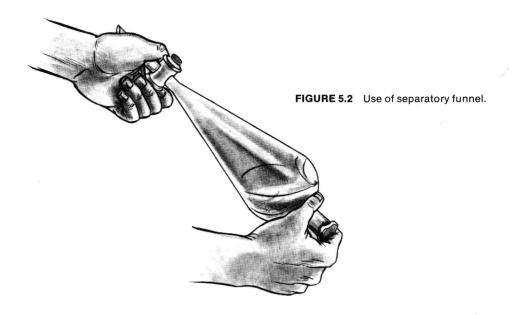

FIGURE 5.2 Use of separatory funnel.

It is sometimes not immediately obvious which layer in the separatory funnel is the organic phase and which is aqueous. If in doubt, withdraw a small sample of the lower phase in a test tube and add a few drops of water to see whether two layers form; if so, the lower phase is the organic solution.

In many cases, the separatory funnel is used simply as a means of re-covering an organic product from a large amount of water with minimum mechanical loss. A small volume of ether or other solvent is added to the mixture to permit sharp separation of layers. Even though the compound may have a negligible solubility in water, a second portion of solvent should be used to rinse the aqueous layer and the separatory funnel.

A common application of extraction is the removal of water-soluble impurities from an organic solution. For example, an ether solution may contain dissolved hydrogen chloride. This is removed by "washing" the ether in a separatory funnel with aqueous carbonate or hydroxide solution and then water.

EXPERIMENT

The objectives of this experiment are to gain experience in extraction, determine by quantitative methods a distribution coefficient and compare the effect of single and multiple extractions. You will first extract an aqueous solution of benzoic acid ($C_6H_5CO_2H$) with methylene chloride (CH_2Cl_2) and determine the amount of acid remaining by titration with standardized base. From this result, the distribution coefficient can be calculated. The effect of extraction with two portions of solvent will then be compared.

Solutions Required. For this experiment each student will use 100 ml of 0.020 M benzoic acid solution and about 20 to 25 ml of dilute sodium hydroxide solution. An additional amount of the acid solution is needed for standardization of the NaOH solution. If students are to prepare the solutions themselves, it is advantageous for pairs of students to collaborate. The directions given provide sufficient amounts for two students; pairs of students can share a buret.

PREPARATION OF 0.020 M BENZOIC ACID SOLUTION. Weigh out 0.610 g (5.00 millimoles or 0.005 moles) of benzoic acid and transfer the acid to a 250 ml Erlenmeyer flask. Add about 150 ml of water and heat over a burner, with swirling, to dissolve the acid. Pour the solution into a 250 ml graduated cylinder or volumetric flask and rinse the flask with several small portions of water to transfer all of the acid. Cool to room temperature and adjust the volume to 250 ml.

PREPARATION OF NaOH SOLUTION. Dissolve 0.5 g of NaOH pellets in 400 ml of water.

1. Standardization of Base

The exact concentration of the NaOH solution is determined by titration of a known volume of the 0.020 M benzoic acid solution. Place 10.0 ml of the acid solution in a small Erlenmeyer flask and add a drop of phenolphthalein. Rinse and fill a buret with the dilute NaOH solution. Record the buret reading and titrate the acid to a permanent pink endpoint. Record the volume and calculate the concentration of base. If you are working in pairs, each student should titrate a sample.

$$\text{millimoles base} = \text{millimoles acid}$$

$$\text{ml base} \times \text{conc. base} \frac{\text{mmoles}}{\text{ml}} = \text{ml acid} \times \text{conc. acid} \frac{\text{mmoles}}{\text{ml}}$$

$$= 10 \times 0.02 = 0.20 \text{ mmoles}$$

$$\text{conc. base} \frac{\text{mmoles}}{\text{ml}} = \frac{0.20 \text{ mmoles}}{\text{ml base required}}$$

2. Distribution Coefficient

Check your separatory funnel with a few milliliters of methylene chloride to see that stopcock and stopper are tight and that the stopcock turns easily (if not, lubricate lightly). Drain the funnel, close the stopcock and add 50 ml of 0.02 M benzoic acid solution. (Measure out 100 ml in a graduated cylinder, pour a little more than half into the funnel and then drain the excess back into the cylinder to the 50 ml mark.) Add 10 ml of methylene chloride (10 ml graduated cylinder), stopper the funnel and extract as described earlier in the chapter. Shake vigorously for at least a full minute. Allow the phases to separate, swirl gently to shake down any droplets of lower layer that cling to the wall, and drain the methylene chloride layer into a small flask. After a minute or two, an additional small amount of lower layer collects and can be drained off.

Drain the aqueous layer into a 250 ml Erlenmeyer flask, rinse the funnel with a few milliliters of water and add the rinse to the flask. Add a drop of phenolphthalein solution, fill the buret with standardized base, and titrate to the same color endpoint used in standardizing the base. Record the volume of base required to neutralize the acid. Continue on to Part 3 and then calculate K_D as outlined in the report sheet.

3. Multiple Extraction

Pour the other 50 ml of 0.02 M benzoic acid solution into the funnel and extract as before using 5 ml of methylene chloride (10 ml graduated cylinder). Drain the methylene chloride layer and extract the aqueous layer again with a second 5 ml portion. Drain the aqueous layer into a flask, add indicator and titrate to determine the amount of acid remaining. Compare with the amount remaining after one extraction with 10 ml of solvent.

4. Calculated Value for Multiple Extraction

From the value of K_D determined in Part 2, the amount of acid extracted by two 5 ml portions can be calculated, as in the example on page 42. Carry out the calculation and compare the calculated value with the value determined experimentally in Part 3.

chromatography

The term chromatography is derived from the original use of this method for separating yellow and green plant pigments. Chromatography has since evolved into a very general separation method for many types of mixtures. It is based on selective **adsorption** of compounds on a solid with high surface area. As the mixture passes over the solid, the components are adsorbed and released from the surface at different rates. They are thus continuously partitioned between the adsorbent and a moving phase, either a vapor or a solution. The process is analogous to separation by fractional distillation or by extraction, in which different compounds are partitioned between liquid and vapor, or between two immiscible liquids, respectively.

Chromatography can be carried out in several ways. In column chromatography, discussed in this chapter, and in thin layer chromatography (TLC), discussed in Chapter 7, a solution of the mixture flows over a solid adsorbent. Separation occurs as molecules are adsorbed and desorbed during passage over the surface. In paper chromatography (Chapter 17), the mixture is partitioned between water molecules adsorbed on the paper and a solvent that moves over the paper. In gas-liquid chromatography (GLC, also called vapor-phase chromatography), a mixture of volatile compounds is separated by passing the vapor over an adsorbent packing in a long, heated tube.

Chromatographic methods have high resolving power, i.e., they are capable of sharp separations of closely related compounds, particularly when very small samples are used. Both GLC and column chromatography can be carried out with instruments that detect extremely small amounts of compounds in the gas stream or liquid, respectively, as it leaves the chromatographic column. In GLC, the detector responds to the thermal conductivity of the gas stream or the ionization of the gas as it passes through a flame. In liquid (column) chromatography instruments, the detector senses changes in the refractive index of the solution. Signals from the detector corresponding to each component in the mixture and proportional to the amount of the compound are recorded automatically on a chart. These instruments thus provide powerful methods for quantitative analysis.

GENERAL PROCEDURE FOR COLUMN CHROMATOGRAPHY

The column of adsorbent, saturated with solvent, is placed in a glass tube with a length about 10 times its diameter. A typical set-up for preparative-scale chromatography is shown in Figure 6.1. A small layer of sand above the adsorbent keeps the surface uniform when liquid is poured in. The most commonly used adsorbents for column chromatography are alumina (Al_2O_3) and silica gel ($SiO_2 \cdot xH_2O$). The solvent used in filling the tube is drained just to the level of the adsorbent, the sample is introduced in a concentrated solution at the top of the column and solvent is then added.

As solvent flows through the column, the chromatogram is **developed,** with the components of the mixture forming bands or zones along the length of the column as they are taken up and released by the adsorbent. If the bands are visible and distinct, the wet column of adsorbent can be pushed out of the tube and cut into the separate zones. More commonly, the solvent flow is continued until the bands are **eluted** or washed off at the bottom of the column and collected as separate fractions.

The choice of solvent in chromatography depends on the nature of the compounds in the mixture. The more polar the compounds, the more strongly they are adsorbed and the more polar the solvent must be to cause them to move at a convenient rate. Useful solvents, in order of increasing polarity, are hexane, benzene, ether, chloroform, alcohols and, finally, water. If the solvent is too polar, all of the compounds will move very close to the "solvent front," and there will be little separation. The chromatogram is developed with the least polar solvent that is effective. A more polar solvent can then be used for the elution.

Two important points must be observed in carrying out column chromatography. The sample must be introduced in a very narrow band at the top of the column. If this is not done, part of the mixture will have been separated, with the less polar components remaining near the top, as more of the mixture flows in. As a result, the bands will spread out and separation will not be achieved. The other precaution is that a solvent level must always be maintained above the adsorbent column; otherwise, channels will form and the bands will become streaked.

EXPERIMENT

The objective of this experiment is to separate the components of a food dye by chromatography. Since the purpose is simply to observe the effect of chromatographic separation, the experiment will be carried out on a micro scale to avoid the need for expensive columns and large volumes of solvents. Quite important also is the fact that because of the small scale, the chromatography can be carried in a very short time, and can be repeated if the results from the first run are not optimal.

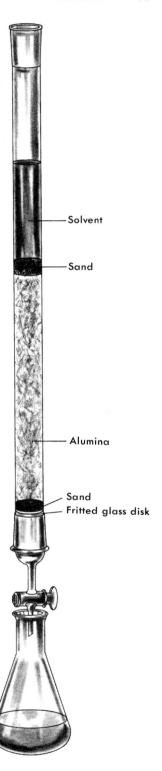

FIGURE 6.1 Chromatography column.

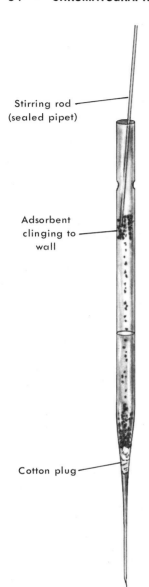

Stirring rod
(sealed pipet)

Adsorbent
clinging to
wall

Cotton plug

FIGURE 6.2 Filling adsorption column.

The tube used for the column is a 2 ml soft-glass pipet. To contain the solvent while the column is prepared, seal off with a small flame the narrow end of a pipet about 1.5 inches below the wide section (see Fig. 6.2). Seal off a second pipet at the tip of the narrow end to form a long slender rod. Place a small loose pellet of cotton in the first tube to support the adsorbent and pack it *gently* into the top of the narrow section with the rod. Mount the column securely in a three-jaw clamp. Obtain about 15 ml of methanol in a small flask. With another pipet, add 1 ml of methanol to the column (half fill); if possible, avoid wetting the walls of the upper end. Wipe the inside of the pipet above the constriction to remove any solvent in the top section.

Obtain 1 g of alumina (20 mm height in a 10 × 75 mm test tube) on a piece of glassine paper, and pour about half of the alumina into the column, using a corner of the paper as a funnel. If the wall is wet, some of the alumina will cling, and the rest will pile up above this point (Fig. 6.2). Stir gently with the rod, scraping with an up-and-down motion to dislodge most of the material, and then add the rest of the alumina. Dislodge again if necessary and then rinse the remaining grains down with a little more methanol. (There should be about 1 inch of solvent above the adsorbent.) Stir the alumina column gently with the rod to eliminate any voids. Record in your report any comments or special notes on the column preparation.

Prepare 10 ml of 80% methanol (8 ml of methanol + 2 ml of water) for use as the developing solvent. In a small test tube obtain a few drops of green or blue dye solution (this is a commercial food coloring, diluted 1:4 with methanol). Have on hand several 1 dram vials for eluate fractions, two pipets and a few milliliters of methanol. Be ready to add 0.1 ml of dye solution (narrow section of pipet half full) when it is time to place the dye on the column.

Read the rest of the procedure before proceeding further, since you will have to keep going after starting the chromatogram. When all preparations are made, gently scratch the tip of the column with a file, snip off the end and allow the methanol to drain into a small beaker. Have the pipet with 0.1 ml of dye ready, and just as the level of solvent in the column reaches the alumina, insert the pipet and add the dye. *Do not let the column go dry;* as soon as the dye has run onto the column, rinse the walls with a few drops of methanol, allow this to drain into the column and then fill to the constriction with methanol. When the level of methanol has nearly reached the alumina, note the appearance of the column and then fill with 80% methanol.

Collect the solvent, adding more 80% methanol as needed to keep liquid above the adsorbent column. Record the approximate amount of solvent used (the free space above the column is about 1 ml). When the first trace of color appears in the eluate, begin collecting in a vial. Continue until there is a change in color in the eluate and then switch to a second vial. When no further color is eluted, add water at the top of the column and continue until all color is eluted or no further change is seen. Record the appearance of the column at various stages, and the color and volume of the eluate fractions.

analysis of drugs by thin layer chromatography

Thin layer chromatography (TLC) is primarily a method for the rapid qualitative analysis of mixtures of organic compounds. As discussed in Chapter 6, a solution of the mixture is passed over an adsorbent surface, and separation occurs as the compounds are adsorbed and released to different extents. The adsorbents and solvents are the same as those used in column chromatography; silica gel is more frequently used as the adsorbent in TLC.

GENERAL PROCEDURE

TLC is carried out on glass plates or strips of plastic coated on one side with a thin layer of adsorbent. The adsorbent contains a small amount of gypsum ($CaSO_4$) which acts as a binder to give an adherent coating. For routine work, small TLC plates can be prepared by dipping microscope slides in a slurry of the adsorbent in chloroform. More uniform plates are obtained by mixing the adsorbent with an equal weight of water, spreading the mixture on a glass plate and allowing it to set and dry. Precoated TLC plates are available commercially with various adsorbents in very uniform layers on a plastic backing.

To "load" the plate, very small samples of the mixture in some volatile solvent are applied as spots near one end and the solvent is allowed to evaporate. The plate is then placed, sample end down, in a closed vessel containing a shallow pool of the developing solvent. The solvent rises on the plate by capillary action, passing over the sample and causing the compounds to move at varying rates depending on their relative affinities for the adsorbent and the solvent. When the solvent film has risen to the top of the plate, the plate is removed and the solvent is allowed to evaporate. The zones or spots containing the various components of the mixture are then detected at various points along the plate. If the compounds are colorless, they are made visible by treating the plate with a reagent, such as iodine vapor, that causes color to develop.

IDENTIFICATION BY TLC

The main uses of TLC are for quick observation of the number of compounds present in a sample, and for qualitative detection of a given compound in a mixture. For the latter purpose, a sample of the compound in question is placed in one position at the bottom of the plate and a sample of the mixture is placed in an adjacent position. With a plate 3 to 4 cm wide, it is possible to place samples in as many as three or four "lanes." If identical compounds are present in two or more lanes, they should appear at the same height after the plate has been developed with solvent. The position of the spot relative to the solvent front, called the R_F value (distance of spot/distance of solvent), depends on the thickness of coating, the amount of sample and the temperature, and may vary from one plate to the next. Comparison of two samples on the same plate is therefore essential.

In addition to the position of the known sample and that of a spot in the mixture, the appearance of the compound on the developed plate may greatly strengthen the identification if it is distinctive after visualization. If there is a reagent specific for the compound of interest, this is sprayed in a fine mist over the surface. A general reagent such as iodine will cause brown or black spots with practically all compounds; these may be slightly different shades, but the color is usually not very characteristic.

Another method for visualizing spots is illumination of the plate with an ultraviolet lamp. Many substances, particularly aromatic compounds, will show a bright fluorescence which may have a characteristic color. A further possibility is use of an adsorbent layer that contains a trace of fluorescent dye. Compounds that are fluorescent still show up as bright spots on a light background; any others appear as a dark spot since they quench the fluorescence of the background dye.

ANALGESIC DRUGS

In this experiment, TLC will be used to examine the composition of various analgesic (pain relieving) drugs. The best known of these is aspirin, but several other chemically similar compounds are also used as analgesics. Among these are phenacetin, salicylamide and acetaminophen. Caffeine (Chapter 9) is sometimes added to these formulations to overcome drowsiness. A few other compounds such as N-cinnamylephedrine (cinnamedrine) and methapyrilene are included for other therapeutic effects, such as antispasmodic or slight sedative action. In addition to the active ingredients, the tablets of these drugs contain starch, lactose and other substances that act as binders and permit rapid solution, and sometimes also inorganic bases.

Aspirin   Phenacetin   Salicylamide   Caffeine

Acetaminophen   Cinnamedrine   Methapyrilene

EXPERIMENT

In this experiment you will obtain as an unknown a proprietary analgesic drug. The objective is to identify the unknown drug by TLC comparison with several known compounds. The unknown will be one of those listed below. The amounts of ingredients in some cases are given in *grains* per tablet [grain (gr) is an apothecary unit; 1 gr = 64 mg].

Drug (Brand name)	Ingredients
Anacin	aspirin and caffeine.
Empirin	phenacetin (2 1/2 gr), aspirin (3 1/2 gr), caffeine (1/2 gr).
Excedrin	acetaminophen (1 1/2 gr), salicylamide, aspirin and caffeine.
Excedrin PM	Acetaminophen (2 1/2 gr), salicylamide, aspirin and methapyrilene hydrochloride.
Midol	aspirin (7 gr), cinnamedrine (0.23 gr), caffeine (0.5 gr).
Tylenol	acetaminophen (325 mg).
Vanquish	aspirin (227 mg), acetaminophen (194 mg), caffeine (33 mg).

The known compounds to be used for reference are the following:

Aspirin (acetylsalicylic acid)

Acetaminophen (4-acetamidophenol)

Caffeine

Phenacetin (p-acetophenetide)

Capillaries for applying the samples can be prepared by heating and drawing out a soft-glass pipet as described on page 13. Alternatively, they can be made from open-end melting point tubes. Soften a 1 cm section in the center of each tube by heating in a low burner flame, remove the tube from the flame and draw it out to a thread-like thickness. Break in the middle of the thin portion to obtain two capillaries.

Label five 1 dram vials with the following designations: 1-Asp, 2-Ace, 3-Unk, 4-Caf, 5-Phen. Place 20 to 30 mg of each of the four knowns in vials 1, 2, 4 and 5. In vial 3 place 1/2 tablet of an unknown analgesic.

To the vials containing the reference compounds, add 0.5 ml of methanol; add 2 ml of methanol to the unknown (or less if a smaller fragment of tablet is used). The concentrations should be in the range 30 to 60 mg/ml; only a small fraction of the solution will be used, and much smaller quantities of sample and solvent can be taken. Crush the unknown tablet gently with a stirring rod and allow the insoluble material to settle. Swirl or stir (use a clean rod) the other samples until all or nearly all of the solid is dissolved.

FIGURE 7.1

Obtain a 7 × 14 cm piece of fluorescent Silica Gel TLC sheet (handle by the edge, and do not touch the coated surface). Samples of the five solutions should be spotted on the coated side about 1/2 inch from one end of the sheet, and about 1/2 inch apart, with the outer two spots about 1/4 inch from the edge of the sheet (Fig. 7.1). The samples should be applied in the order 1 to 5 from left to right, with the unknown in the middle lane. It is most important to avoid applying too large an amount of sample. The spot after application should be about 1 to 2 mm (1/16 inch) in diameter. It is a good idea first to practice applying spots to a small scrap of sheet.

To apply the samples, touch the end of a capillary tube to the solution and then touch this gently to the plate at the proper place. When the five samples have been applied, place the sheet, spotted end down, in a developing jar containing a pool of solvent about 1/4 inch deep. The solvent system used in this experiment is a mixture of benzene (120 ml), ether (60 ml), acetic acid (18 ml) and methanol (1 ml). Cap the jar securely and develop the chromatogram. About 40 minutes are required for the solvent to rise to within 1 to 2 cm of the top of the sheet.

At this point, remove the sheet, and mark the solvent boundary with a small scratch. Recap the jar and allow the sheet to dry. Examine the chromatogram under a UV lamp and sketch in your report the appearance of the plate, indicating the location and approximate size of the spots and any distinctive colors. After this examination, place the sheet in a jar of iodine vapor for about 30 seconds, remove and again record the appearance.

steam distillation
of clove oil

In the distillation of a mixture (more precisely a solution) of two miscible liquids, the composition of the distillate depends on the amounts and the vapor pressures of the two components (Fig. 4.2). A different situation is encountered in the distillation of a mixture of two compounds that are not mutually soluble. In this case, the total vapor pressure above the mixture is simply the sum of the individual vapor pressures, $P_T = pA + pB$, *independent* of the amounts of the two compounds. As long as some of each liquid phase is present, the distillate will have a constant composition and the boiling point will be *lower* than that of either compound alone because the vapor pressures of both compounds contribute to the total vapor pressure required for distillation.

When one of the two components is water, this process is termed steam distillation, and it provides a means of distilling slightly volatile compounds at a temperature far below the atmospheric boiling point. Steam distillation is useful in the separation of any organic compound that is present in a large amount of nonvolatile material. In this case, simple distillation, even in vacuum, is ruled out because destructively high temperatures would be needed to distill the compound from a porous mass. Steam distillation is a useful alternative to extraction for the isolation of volatile organic compounds, such as essential oils from plant material. Extraction with solvents removes gums and fats as well as the volatile oils; the latter are separated selectively by steam distillation.

Steam distillation is conveniently carried out in a two- or three-neck flask to permit introduction of water or steam (Fig. 8.1). The use of steam provides necessary agitation when a bulky or tarry mass is being steam distilled. Addition of water with external heating is more convenient for small scale work; this procedure is used in the present experiment.

ESSENTIAL OILS

The characteristic aromas of plants are due to the volatile or essential oils, which have been used since antiquity as a source of fragrance and

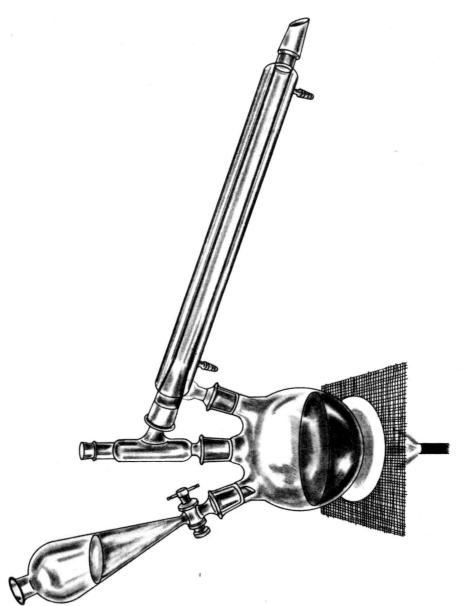

FIGURE 8.1 Steam distillation set-up.

flavoring. These oils occur in all living parts of the plant; they are often concentrated in twigs, flowers and seeds. Essential oils are generally complex mixtures of hydrocarbons, alcohols and carbonyl compounds, mostly belonging to the broad group of plant products known as terpenes. Some terpenes and a few hydrocarbons are found in many different plant species. Certain other compounds, particularly aromatic aldehydes and phenols, are a major constituent in the essential oils of one or a few plants, and impart the characteristic aroma of condiments such as cloves, cinnamon and vanilla.

A good example of steam distillation is the isolation of the essential oil from cloves *(Eugenia caryophyllata)*. The pleasant smelling oil consists mainly of 4-allyl-2-methoxyphenol, or eugenol; this compound is familiar as a flavoring and also as a mild local analgesic in dentistry.

EXPERIMENT

The objective of this experiment is to get experience in carrying out steam distillation by isolating eugenol from cloves.

Clamp a 500 ml three-neck flask on a ring and gauze above a Bunsen burner, and attach a side-arm adapter and condenser to the center neck in the usual manner for simple distillation. In one of the side necks place a separatory funnel. If a 250 ml funnel (without ground glass joint) is used, fit the stem of the separatory funnel into the neck of the flask with a one-hole cork or rubber stopper and support the weight of the funnel with a clamp or ring. If a separatory funnel with joint is available, simply place this in the neck.

Weigh out 12 g of ground cloves (or use a volume designated by the instructor) and place in the flask. Add 100 ml of water and stopper the other neck of the flask and the top of the side-arm adapter. Use a 125 ml Erlenmeyer flask as a receiver. Place about 100 ml of water in the separatory funnel. Turn on condenser water.

Heat the contents of the flask to boiling and distill at a steady rate. Add water periodically from the separatory funnel to maintain the original liquid level in the flask. After about 100 ml of distillate have been collected, change receivers, and if oil droplets are still collecting in the receiver, distill over another 20 ml of water.

Pour the distillate into separatory funnel, rinse the receiver with 5 ml of methylene chloride, and add this to the funnel. Shake the funnel, allow the layers to separate cleanly and drain the organic layer into an 18 × 150 mm test tube which has been previously weighed (this is the *tare* weight). Rinse the aqueous layer and separatory funnel with 2 ml of methylene chloride and drain carefully into the test tube. If the separation has been made cleanly, drying of the solution is unnecessary. If there is a significant amount of water present, add a little sodium sulfate, swirl and decant the solution through a cotton wad into another tared test tube.

Concentrate the methylene chloride on the steam bath, shaking to prevent boiling over. When the solvent is essentially all evaporated, remove the last traces by tipping the tube to spread the oily residue around the lower part of the wall and directing a *brief, gentle,* stream of air into the tube.

Weigh the tube and residual eugenol and calculate the yield, based on the weight of cloves used.

REPORT: CHAPTER 8

1. Description of Distillation

(Give a concise account of your actual experience and observations in the distillation; include appearance of distillate, total volume distilled and any incidents that might affect the outcome of the experiment.)

2. Isolation of Product

(Give a brief description—was the solution dried before evaporation?)

3. Product

Weight of test tube + eugenol:

Tare weight of test tube:

Weight of eugenol:

% Yield based on cloves used:

QUESTIONS

1. Suggest another possible method that might be used to obtain eugenol from cloves.

2. What is the purpose of the water (steam) in the procedure used? (Why not simply distill out the eugenol directly without steam?)

3. Could methylene chloride (or a higher boiling organic compound with similar solvent properties) be used instead of water for the codistillation of eugenol from cloves? Explain your answer.

4. Since the pressure of a gas is proportional to the number of moles at a given temperature and volume, the partial pressures P_{H_2O} and P_{eug} in the steam distillation are proportional to the numbers of moles (n) of water and of eugenol, respectively, or the weights divided by molecular weights:

$$\frac{P_{H_2O}}{P_{eug}} = \frac{n_{H_2O}}{n_{eug}} = \frac{\text{weight } H_2O/18}{\text{weight eugenol}/164} \text{ or}$$

$$\frac{\text{weight } H_2O}{\text{weight eugenol}} = \frac{P_{H_2O}}{P_{eug}} \times \frac{18}{164}$$

Assume 755 as the atmospheric pressure during the experiment. This is the total pressure P_T that must be equaled by the sum of the partial pressures $P_{H_2O} + P_{eug}$. In the range 96 to 100°, the vapor pressure of eugenol is about 8 mm. Thus, the vapor pressure of water during the distillation must be

$$P_{H_2O} = P_T - P_{eug} = 755 - 8 = 747 \text{ mm}$$

a. From Figure 4.1 (page 32), estimate the temperature at which the distillation occurred.

b. How many g of water are required to effect the steam distillation of 1 g of eugenol?

isolation of caffeine

Caffeine is a compound present in the fruit and bark of a number of plants, including tea, coffee, cacao and cola nuts. The caffeine content is 3 to 4% in dried tea leaves and 1.2% in green coffee beans. Caffeine is a mild stimulant and also has a diuretic action; it is widely used in proprietary drugs for the stimulant effect.

Caffeine is a derivative of the very important group of compounds called **purines,** which are among the major components of nucleic acids. The purine nucleus is found in a number of forms in naturally occurring compounds, including **uric acid,** which is the form in which nitrogen is excreted in nonmammalian animals. The presence of methyl groups in caffeine is a result of biochemical methylation, a common process in plant metabolism.

Caffeine   Uric acid   Purine

The extraction of caffeine from coffee is an important commercial process, since the effects of the drug are considered undesirable by some persons. Decaffeination is carried out by treating the green coffee beans with a small amount of hot water and then by exposing to a solvent until 97% of the caffeine has been removed. The solvent used is trichloroethylene; residual solvent is removed after extraction by steam distillation. The process also removes wax from the beans, which are then roasted in the usual way.

The solubility of caffeine at room temperature is 2.2 g/100 ml in water and 18 g/100 ml in chloroform. The compound crystallizes readily and it

79

can also be sublimed. In the laboratory isolation of caffeine, use is made of all of these properties. The caffeine is first extracted from tea leaves or ground coffee by hot water. This step also removes tannins and other materials; these are precipitated by addition of a metal carbonate, which converts tannins and acids to insoluble salts. The caffeine is then recovered from the aqueous solution by liquid-liquid extraction with chloroform. After removing water by drying the chloroform solution, the solvent is evaporated and the crude caffeine is purified by sublimation.

DRYING AGENTS

After extraction of an aqueous solution, the organic phase is saturated with water, and it is desirable to dry the organic solution before evaporating the solvent. Water is an impurity, and it must be removed before carrying out crystallization or distillation of the organic compound that is being isolated. A number of anhydrous salts can be used for this purpose; the most common ones are magnesium or sodium sulfate. For solutions, the amount of drying agent should usually be about one-tenth of the liquid volume; a smaller amount is used for drying a pure liquid. After standing over the drying agent for 10 to 20 minutes, with occasional shaking, the solution is filtered through a cotton plug and the drying agent is rinsed with solvent.

SUBLIMATION

Sublimation is the process of vaporizing and condensing a solid. A solid compound has a vapor pressure, and although this may be low, vaporization can occur from a solid just as from a liquid. With a few compounds that have high melting points, the vapor pressure may reach 760 mm at a temperature below the melting point, and no liquid state exists at atmospheric pressure.

Sublimation is an alternative to recrystallization for purification of a solid compound, and it is useful when the compound is contaminated with impurities of high molecular weight. It is not a useful procedure for separation of closely related compounds. Vaporization of a solid occurs from a dry surface without boiling or bubbling, and a column for conducting the vapor to the condenser and returning condensate to the flask is not needed as it is in a distillation. The condensing surface is held quite close to the material being sublimed; it is usually a narrow "cold finger" so that the sublimed material is collected on a small area. A simple apparatus for sublimation is shown in Figure 9.1. After the sublimation is complete, the solid is scraped from the cold finger with a spatula.

FIGURE 9.1 Sublimation apparatus.

Ice and water

EXPERIMENT

The objective of this experiment is to isolate caffeine by carrying out the extraction of tea or coffee and purifying the product by sublimation.

Initial Isolation

From Tea. Weigh out 15 g (50 ml) of tea leaves and 15 g of powdered calcium carbonate, and place these in a 250 ml Erlenmeyer flask. (The calcium carbonate forms an insoluble precipitate with tannic acid.) Add 150 ml of water and heat the mixture to boiling over a burner. Boil gently for 15 to 20 minutes; stir occasionally and make sure that the mixture does not froth over.

From Coffee. Weigh out 30 g (75 ml) of ground coffee, place it in a 250 ml flask with 150 ml of water and boil for 10 minutes. Allow the mixture to cool slightly, add 5 g of lead carbonate to the flask and boil for an additional 5 minutes. (The lead salt is usually used for precipitation of the tannins from coffee; it has the same purpose as the $CaCO_3$ used in the tea procedure.)

Further Steps for Either Source of Caffeine

While the tea or coffee is brewing, prepare a suction filter using a 500 ml side-arm filter flask and a 3 or 4 inch Büchner funnel with the proper size of filter paper. Moisten the paper and apply suction briefly with the aspirator to seat the paper. If you are using coffee or ground tea, cover the paper to a depth of about 1/8 to 1/4″ with Celite or filter aid. These are finely divided porous mineral materials which are used to prevent clogging of the filter paper; if the filtration becomes slow, the filter aid pad can be scraped to expose fresh porous surface. (With whole tea leaves, the Celite is unnecessary and interferes with squeezing the liquid from the bulky mass.)

When the aqueous extraction is complete, cool the mixture somewhat to allow easier handling, then turn on the aspirator and pour the mixture into the funnel. Keep one hand on the tubing to the aspirator and pinch off or kink the tubing as needed to prevent the filtrate from frothing up into the side-arm. After the liquid has been transferred, shake as much of the solid as possible into the funnel, rinse the flask with 10 ml of water and pour this into the filter. Squeeze the grounds or leaves and then discard the filter cake into a waste can, *NOT* IN THE SINK.

Cool the tea or coffee extract to room temperature and pour it into a separatory funnel (stopcock closed!). Obtain 75 ml of chloroform in a 100 ml cylinder and pour 20 ml of this into the separatory funnel. The extraction is now carried out as described on page 42. Do not shake the mixture vigorously, or a very stable emulsion will be formed because of the presence of surface-active compounds in the extracts. Mix the two layers thoroughly by gently shaking them together for 1 minute and then allow the layers to separate for about 4 to 5 minutes. The bottom chloroform layer should have a definite yellow color. This layer will usually still be partially emulsified, but it does not pay to wait more than 5 minutes. Drain the bottom layer slowly into a 125 ml Erlenmeyer flask, stopping before the darker-appearing upper layer reaches the stopcock. Repeat the extraction twice with two further 20 ml portions of chloroform, draining each chloroform layer into the same flask.

After the third extraction, discard the aqueous layer from the funnel into the sink. The emulsion in the chloroform extract can usually be substantially reduced by suction filtration. Put a clean paper in the Büchner funnel, rinse the 500 ml filter flask with water and attach the aspirator. Pour the combined chloroform layer through the filter and transfer the filtrate again to the separatory funnel. Rinse the funnel and filter flask with 5 ml of chloroform and add this to the main solution. At this point, there will be a small amount of emulsion at the top of the chloroform; this contains some chloroform and caffeine, but it does not pay to try to recover it. (If the chloroform layer is still very heavily emulsified, consult your instructor before proceeding further.) Drain the clear (or slightly cloudy) chloroform solution into a dry 125 ml Erlenmeyer flask and add about 1 teaspoonful of magnesium sulfate. Cork the flask and allow it to stand for 5 to 10 minutes with occasional shaking.

Clamp a 100 ml round-bottom flask in position for distillation, set a glass funnel above it on a ring and place a pea-sized ball of cotton into the bottom of the funnel. Press the cotton slightly into the stem of the funnel, but do not wad it into a tight pellet. Pour the chloroform solution through the cotton and rinse the drying agent and flask with a little more chloroform.

Add a boiling stone to the solution, and set up for simple distillation (Fig. 4.4) with a distillation head, condenser and adapter; use a 125 ml Erlenmeyer flask for a receiver. A cork can be used instead of the collar and thermometer to close the top of the distillation head. Distill off the chloroform until the volume of the solution in the flask is about 5 ml – do not allow the contents to go dry. The distillation can be carried out as rapidly as the condenser will permit (the distillate should not be warm).

Sublimation

The residual caffeine solution is now transferred to a small filter flask for sublimation. Using a pipet and rubber bulb, draw up the concentrated solution from the distilling flask and transfer it to a 125 ml side-arm flask. Rinse with a little solvent to obtain a complete transfer, but keep the volume as small as possible. Clean and put away the distillation apparatus, and return the chloroform distillate to a bottle on the side shelf.

Clamp the filter flask containing the caffeine concentrate to a ring stand, attach the side-arm to the aspirator with heavy-wall tubing, and stopper the flask with a solid rubber stopper. Turn on the aspirator and allow the residual chloroform to evaporate. Warm the flask with your hand and swirl gently until the residue is a solid crust on the bottom of the flask.

For sublimation, disconnect the aspirator tubing and replace the rubber stopper with a cold finger (Fig. 9.1). This is a test tube fitted with a rubber collar (return to side shelf after use). If necessary, adjust the height of the collar so that the bottom of the test tube is about 1/2″ from the bottom of the flask. Fill the test tube to a depth of about 1″ with small pieces of ice, attach the aspirator and evacuate the flask. If the cold finger slips further down to the bottom of the flask, open the system and readjust the height.

To sublime the caffeine, heat the bottom of the flask with a low flame, starting at the outer edge and moving in. As the sublimation proceeds, some of the caffeine may be deposited on the cooler wall of the flask; it can be resublimed to the cold finger by heating the wall. If the ice in the cold finger melts completely, remove some of the water with a pipet and add a little more ice. Continue until all of the solid sublimes, leaving only a dark film in the flask.

Release the vacuum and carefully remove the cold finger. Scrape the caffeine onto a piece of glassine paper, weight it and submit it to your instructor in a vial labeled with your name, the weight of the product and the source (coffee or tea). Calculate the yield of caffeine (percentage weight of the original tea or coffee).

reactivity in
substitution reactions

Alkyl halides or alkyl sulfonate esters are the reactants or "substrates" in many nucleophilic substitution reactions. Two typical substitutions are illustrated by the reactions of s-1-chloro-1-phenyl-ethane with cyanide ion and with water. In both reactions, a nucleophile is substituted for the chlorine, but the reactions occur by different mechanisms, called S_N2 (substitution, nucleophilic, bimolecular) and S_N1 (substitution, nucleophilic, unimolecular).

As shown in the upper equation, the S_N2 reaction involves the simultaneous bonding of the CN^- ion at the rear side of the asymmetric carbon as the Cl^- leaves. Since both cyanide ion and chloro compound participate in the transition state, the rate of the reaction depends on both species. The configuration of the asymmetric center is *inverted*, i.e., s-chloride gives R-cyanide. The S_N1 process occurs by dissociation of the chloride to give a carbocation which is planar. Reaction with water therefore gives equal amounts of the R- and s-alcohol. The rate is independent of the concentration of the nucleophile because the slow step in this case is the breaking of the C—Cl bond.

In a given substitution reaction such as the reaction of a halide with sodium ethoxide in ethanol to give the ether, the product may be formed in part by both mechanisms, operating concurrently. Reaction by one path or by

$$R—X + NaOC_2H_5 \xrightarrow{C_2H_5OH} R—OC_2H_5 + NaX$$

the other can be examined, however, by the proper choice of nucleophile and reaction conditions. The S_N2 process requires a good nucleophile, such as cyanide, thiolate or iodide anions. The S_N1 mechanism requires a good ionizing solvent, such as water, for the dissociation step.

To compare the reactivity of a series of alkyl halides in the two mechanisms, the S_N2 reaction can be observed by the displacement of chlorine or bromine by iodide ion in acetone solution. Acetone is a relatively poor ionizing solvent, and S_N1 dissociation is minimized. Sodium iodide is very soluble in acetone, but sodium chloride and sodium bromide have very low solubilities. The course of the reactions can be seen by the formation of a crystalline deposit of NaCl or NaBr.

$$R—X + NaI \xrightarrow{acetone} R—I + NaX \downarrow$$

The S_N1 reaction can be observed by treating the alkyl halide with a solution of silver nitrate in aqueous ethanol. Nitrate ion is a very poor nucleophile, and there is little opportunity for S_N2 displacement. Dissociation of the alkyl halide by the S_N1 process is seen by the precipitation of the insoluble silver halide; the carbocation is then captured by alcohol or water.

$$R—X \longrightarrow R^+ + X^- \xrightarrow{AgNO_3} AgX + NO_3^-$$

Alkyl halides vary widely in their reactivity in the two substitution processes. For the S_N2 displacement, the nucleophile must be able to attach from the rear of the leaving group. **Steric factors** are therefore very important; the more room available for the incoming nucleophile, the lower the energy barrier and the faster the reaction. In the S_N1 reaction, the **stability of the carbocation** is the decisive factor. The better stabilized the cation, the faster the reaction.

EXPERIMENT

The objective of this experiment is to observe the reactions of a series of alkyl halides with solutions of NaI in acetone and with AgNO$_3$ in aqueous ethanol, and to compare the reactivities as a function of structure. The five halides used are:

1. $CH_3CH_2CH_2CH_2Cl$ n-butyl chloride

2. $CH_3CH_2CH_2CH_2Br$ n-butyl bromide

$$\overset{\text{Cl}}{\underset{|}{}}$$

3. $CH_3\overset{|}{C}HCH_2CH_3$ *sec*-butyl chloride

$$\overset{\text{CH}_3}{\underset{|}{}}$$

4. $CH_3—\overset{|}{\underset{|}{C}}—Cl$ *tert*-butyl chloride

$$\underset{\text{CH}_3}{}$$

5. $CH_3CH{=}CHCH_2Cl$ crotyl chloride

Label two series of 5 clean, dry test tubes from 1 to 5. In each series of tubes, place 0.2 ml of the following halides: tube 1: *n*-butyl chloride; tube 2: *n*-butyl bromide; tube 3: *sec*-butyl chloride; tube 4: *tert*-butyl chloride; tube 5: crotyl chloride. Keep the tubes stoppered with corks; leave the corks in at all times, before and after adding reagents. Obtain 15 ml of 15% NaI-acetone solution and 15 ml of 1% ethanolic $AgNO_3$ solution from the side shelf.

Arrange one series of tubes in order from 1 to 5. Add 2 ml of the **NaI solution** to tube number 1 and note the time. (Add the solution from a pipet as rapidly as possible—not dropwise.) After 2 to 3 minutes, add 2 ml of NaI solution to the second tube and again note the time. Continue at 2 to 3 minute intervals with the remaining tubes. After each addition, watch for any rapid reaction and then inspect the other tubes for signs of a precipitate. Note the time as closely as possible when precipitation begins to occur, recording the data in tabular form in your notebook. Cork the tubes and allow them to stand, observing them periodically while the next series is run.

Arrange the second series of tubes, and in the same way add 2 ml portions of the **$AgNO_3$ solution** to each tube at 2 minute intervals. Again, watch closely for any rapid changes and then observe the others periodically. If possible, note the time both for the first appreciable turbidity and also for a definite precipitate. If any tubes in the NaI series are still clear at this point, loosen the corks slightly and place the tubes in a water bath at 50°; note any changes that occur.

When you have completed these test tube reactions, pour the contents of all of the test tubes into the container in the hood labeled "waste solvent."

reactions of alcohols

Alcohols are the most readily available and versatile class of organic compounds. A number of alcohols are commercially available in large quantities, since they are starting materials for plasticizers, coatings, surface-active agents and other large-scale applications. These industrial alcohols include all those with up to four carbons, many C_5 and C_6 alcohols, and the *n*-alkanols with even-numbered chains through C_{18}. Other alcohols with any desired carbon skeleton can be obtained on a laboratory scale by reactions of Grignard reagents with carbonyl compounds.

Alcohols undergo elimination reactions to give alkenes, displacements to give alkyl halides, oxidation to aldehydes and ketones, carbonyl condensation to give acetals and acylation to give esters.

The generality and usefulness of these reactions depend on whether the alcohol is **primary** (RCH_2OH), **secondary** (R_2CHOH) or **tertiary** (R_3COH). In reactions that involve breaking the C—O bond to form an electrophilic carbon, as in dehydration or substitution by halide, the reactivity order is $3° > 2° > 1°$. When the reaction involves the hydroxyl group as a nucleophile, with breaking of the O—H bond, the reactivity order is *reversed*, and reactions with *tert*-alcohols are very slow.

Dehydration or Substitution: $3° > 2° > 1°$

Esterification: $1° > 2° > 3°$

DEHYDRATION

The first step in the elimination of water from an alcohol is the transfer of a proton from a strong acid such as H_2SO_4 to the hydroxyl group. Loss of water then occurs to give a carbocation, and rapid removal of a proton leads to the alkene. Tertiary alcohols are most reactive, and primary alcohols least, because the *ease of formation and stability of carbocations increases with the number of alkyl groups attached to the electron-deficient carbon.*

| 3° carbocation most stable | 2° carbocation | 1° carbocation least stable |

Dehydration of a primary alcohol with sulfuric acid requires 80 to 90% H_2SO_4 and a temperature of 140° or higher. Secondary and tertiary alcohols can be dehydrated under progressively less vigorous conditions.

When a tertiary alcohol is treated with *concentrated* H_2SO_4 at a low temperature, a further reaction can occur. Under these conditions the tertiary cation can exist for sufficient time to permit reaction with a molecule of alkene. A new cation is then formed and loss of a proton gives a _dimeric_ alkene. With _tert_-butyl alcohol, the process is as follows:

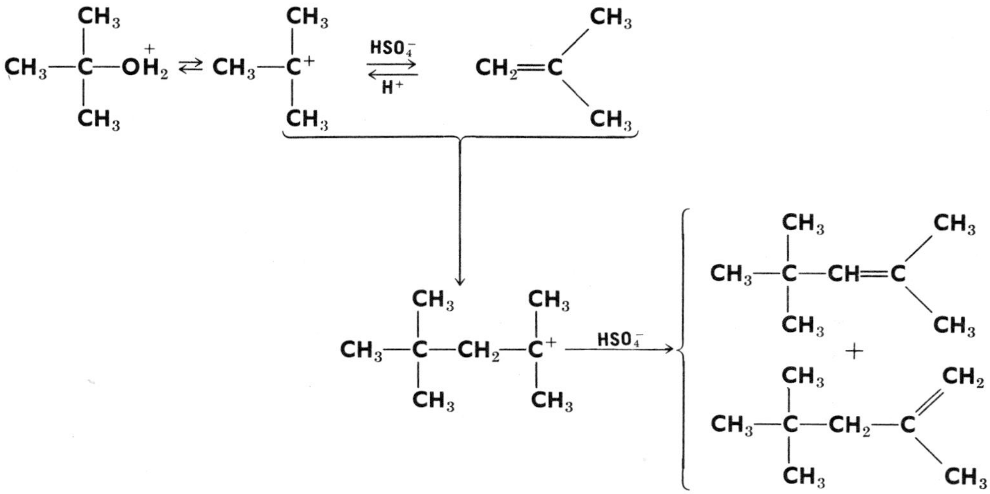

SUBSTITUTION

In substitution reactions of alcohols, just as in elimination, the —OH group must be protonated so that a water molecule can be displaced. This means that the reactions must be carried out under strongly acid conditions. The nucleophiles that can be used in substitutions are therefore limited to those that are not protonated in strong acid. Halide ions meet this requirement, and the most useful substitution reactions of alcohols are the formation of alkyl chlorides and bromides.

$$R\text{—}OH + H^+X^- \longrightarrow R\text{—}\overset{+}{O}H_2 \quad \text{:}X^- \longrightarrow RX + H_2O$$

Preparation of the chloride from a tertiary alcohol can be carried out simply by mixing the alcohol with concentrated hydrochloric acid. Secondary alcohols react very slowly with HCl, but in the presence of zinc chloride, which acts as a Lewis acid, secondary alcohols are converted to the chloride at room temperature. A solution of zinc chloride in HCl, called the Lucas reagent, can be used to distinguish primary, secondary and tertiary alcohols containing up to six carbons. These alcohols dissolve in the reagent, and the time required for separation of a separate layer of the alkyl chloride is an indication of the type of alcohol.

OXIDATION

Oxidation of primary and secondary alcohols is a general method for obtaining carbonyl compounds. One of the most useful reagents is **chromic acid,** which can be used in the form of chromic anhydride, CrO_3, or sodium dichromate, $Na_2Cr_2O_7$, in sulfuric acid. The reaction occurs by formation of a chromate ester and elimination of the chromate group in a +4 valence state. The chromium ends up as the green Cr^{+3} ion.

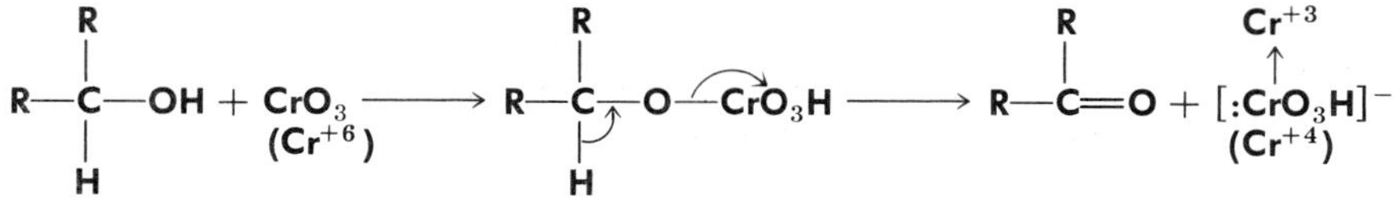

$$R-\underset{\underset{H}{|}}{\overset{\overset{R}{|}}{C}}-OH + CrO_3 \ (Cr^{+6}) \longrightarrow R-\underset{\underset{H}{|}}{\overset{\overset{R}{|}}{C}}-O-CrO_3H \longrightarrow R-\overset{\overset{R}{|}}{C}=O + [:CrO_3H]^- \ (Cr^{+4}) \quad Cr^{+3}$$

The oxidation of a primary alcohol gives an aldehyde, but further oxidation to the carboxylic acid takes place very readily. Secondary alcohols are oxidized to ketones in high yield. A tertiary alcohol can react only by breaking a C—C bond, and is unaffected by cold chromic acid. The three types of alcohols can be differentiated by their reactions with chromic acid. A color change from bright orange to dark green indicates oxidation of a primary or secondary alcohol. If the oxidation product is a ketone, resulting from a secondary alcohol, treatment with 2,4-dinitrophenylhydrazine gives a crystalline yellow hydrazone. The acid resulting from complete oxidation of a primary alcohol does not react with 2,4-dinitrophenylhydrazine. The intermediate aldehyde does form a hydrazone, however, and this can be obtained if the oxidation is incomplete.

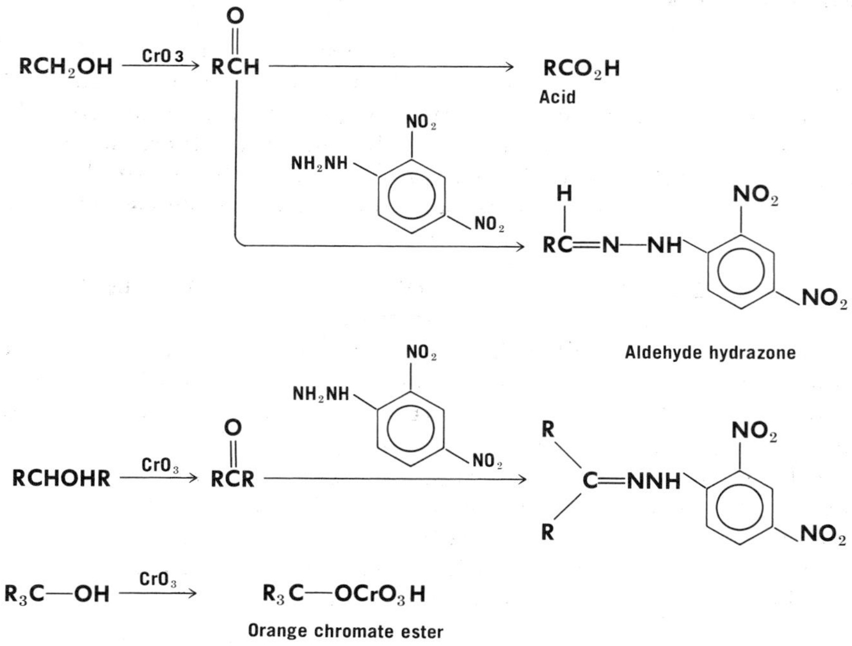

Aldehyde hydrazone

Orange chromate ester

EXPERIMENTS

The objective of the experiments in this chapter is to observe and compare the behavior of three alcohols in several characteristic reactions. The alcohols used are as follows:

n-butyl alcohol $CH_3—CH_2—CH_2CH_2OH$ b.p. 117°

sec-butyl alcohol $CH_3—CH_2—\overset{\overset{\displaystyle OH}{|}}{CH}—CH_3$ b.p. 99°

tert-butyl alcohol $CH_3—\overset{\overset{\displaystyle OH}{|}}{\underset{\underset{\displaystyle CH_3}{|}}{C}}—CH_3$ b.p. 83°

Dehydration

The four-carbon alkenes have boiling points below room temperature, and evidence of a reaction is the formation of a gas which undergoes addition of bromine. Set up a water bath in a large beaker over a burner, with a clamp mounted just above the bath. Obtain a 16″ length of rubber or plastic tubing with a 6 mm glass connecting tube at each end. Insert one connector through the larger end of a one-hole rubber stopper which fits into an 18 × 150 mm test tube (lubricate with glycerine before inserting tubing!). Label six test tubes, two "n", two "sec" and two "tert," and in each tube place 2 ml of the corresponding alcohol. One set of the three alcohols will be treated with 30% H_2SO_4 and the second with concentrated H_2SO_4 (95%). Obtain from the side shelf about 15 ml of 2% Br_2—CCl_4 solution, 5 ml each of concentrated H_2SO_4 (CAUTION!) and 30% H_2SO_4, and a few boiling stones. Heat the water bath to about 70°.

30% H_2SO_4. To each of the alcohols in one set of tubes add 1 ml of 30% H_2SO_4. Afer adding the acid, shake each tube to mix and note any changes. Starting with the *tert* alcohol, add a boiling stone and clamp the test tube in the water bath so that the bottom third is immersed. Put the stopper connected to the tubing in place and put the other end of the tubing into a test tube containing 2 ml of the bromine solution. Heat the bath until steady bubbling occurs and continue for 3 minutes or until the red bromine color disappears. Note the time required to decolorize the bromine.

Turn off the burner, remove the test tube and replace it with the test tube containing the *sec* alcohol. Add a boiling stone, put 2 ml of Br_2—CCl_4 in the trap, connect the tubing and again heat. Raise the bath temperature to gentle boiling and continue for the same time used for the *tert* alcohol. Note any gas evolution or changes in the bromine color.

If dehydration (evolution of alkene) is observed with the secondary alcohol, repeat the process with the primary alcohol, following the same procedure (boiling stone, fresh Br_2 solution and so forth). If no reaction was observed for the secondary alcohol, omit the test of the *n*-butanol.

Concentrated H_2SO_4. Chill the other set of tubes in an ice bath and then, beginning with the *primary* alcohol, add 1 ml of concentrated H_2SO_4. After each addition, shake the tube to mix the contents and note any changes. Proceed to the secondary and then the tertiary alcohol. If any of the test tubes contains two liquid layers, remove a sample of the upper layer with a pipet and add it to 1 ml of Br_2—CCl_4 solution.

Heat the three samples in the water bath, following the same procedure used for the samples with 30% H_2SO_4. Add a boiling chip to each test tube, connect it to the test tube containing Br_2—CCl_4 solution and heat at 90 to 95° for 3 minutes or until 2 ml of Br_2—CCl_4 is decolorized. Record your observations and compare the reactivities of the three alcohols under the two sets of conditions.

Substitution

Place 2 ml of Lucas reagent in each of three small test tubes. (The reagent is prepared by dissolving 136 g of zinc chloride in 80 ml of concentrated hydrochloric acid.) To one tube add 0.2 ml of *n*-butanol, to another add 0.2 ml of *sec*-butanol and to the third add 0.2 ml of *tert*-butanol. (Fill narrow section of soft glass pipet to obtain 0.2 ml.) After each alcohol is added, shake the tube and watch for any indication of a reaction. If no reaction is noticeable immediately after mixing, observe the tubes after 1, 2, 5 and 15 minutes and record any changes.

Oxidation

Place 0.2 ml of *n*-butanol in an 18 × 150 ml test tube, 0.2 ml of *sec*-butanol in a second tube and 0.2 ml of *tert*-butanol in a third tube. To each tube add 0.2 ml of water and shake the tubes to mix the contents. Obtain about 5 ml of chromic acid solution (prepared by dissolving 25 g of CrO_3 in 75 ml of water plus 25 ml of concentrated H_2SO_4).

(CAUTION! The oxidation of an alcohol, even on a small scale, is a very exothermic reaction. Do not "scale up" the amounts specified. When adding the chromic acid solution, be sure to mix the reactants thoroughly after each drop of acid is added.)

Start with the tertiary alcohol and add, *drop by drop*, with a pipet and bulb, 0.2 ml of chromic acid solution. Shake the tube to mix after each drop is added. Note whether the mixture becomes warm.

In the same manner, add 0.2 ml of chromic acid to the secondary alcohol. Shake the tube after each drop is added, and cool in a water bath if the mixture becomes appreciably warm. After 2 minutes, record the color and add, dropwise, another 0.2 ml of acid.

Add 0.2 ml of chromic acid to the primary alcohol, mixing and cooling after each drop. After 2 minutes add another 0.2 ml of chromic acid, dropwise with cooling. At this point, remove about 0.2 ml of the reaction mixture with a pipet and put it in a small test tube. To the remaining mixture add two further 0.2 ml portions of chromic acid. Note the color and the odor of the mixture.

To the oxidation mixtures from the secondary and primary alcohols and the sample removed from the primary alcohol reaction, add 1 to 2 ml of hexane. Shake the tubes to mix the contents thoroughly and pour or pipet off the hexane layers into three small test tubes that have been thoroughly rinsed with water (not acetone!). To each of the hexane solutions add several drops of 2,4-dinitrophenylhydrazine solution, shake the mixtures and record your observations.

aromatic substitution

ELECTROPHILIC AROMATIC SUBSTITUTION

The most common and important reactions of aromatic compounds are **electrophilic substitutions,** in which a group E replaces hydrogen on the aromatic ring. These reactions occur by attack of an electrophilic reagent E^+ on the aromatic ring to give an intermediate cation, followed by loss of a proton. When a substituent is present that can *donate electrons* and stabilize the intermediate, the incoming electrophile is directed to the *ortho* and *para* positions. Chlorine is an example of an *o,p*-directing group.

Nitration is a typical electrophilic substitution. The electrophile is the nitronium ion NO_2^+. This ion is usually generated by carrying out the reaction with nitric acid in the presence of sulfuric acid. In this mixture, nitric acid acts as a *base,* and after it is protonated, water is lost. With a large excess of sulfuric acid, nearly all of the HNO_3 is converted to the NO_2^+ ion.

$$HO-NO_2 + H-OSO_3H \rightleftarrows H_2\overset{+}{O}-NO_2 + HSO_4^-$$

Nitric   Sulfuric
acid   acid

$$H_2\overset{+}{O}-NO_2 \rightleftarrows H_2O + NO_2^+$$

$$H_2O + H_2SO_4 \rightleftarrows H_3O^+ + HSO_4^-$$

Overall: $HNO_3 + 2\ H_2SO_4 \rightleftarrows NO_2^+ + H_3O^+ + 2\ HSO_4^-$

In the nitration of chlorobenzene, two *ortho* and one *para* positions are available for substitution. Stabilization of the intermediate is somewhat greater for the *para* position, and the *para* isomer predominates. After one nitro group is introduced, the ring is much less susceptible to further reaction since the NO_2 group is *electron-withdrawing* and therefore deactivates the ring for further electrophilic attack, particularly at the *ortho* and *para* positions.

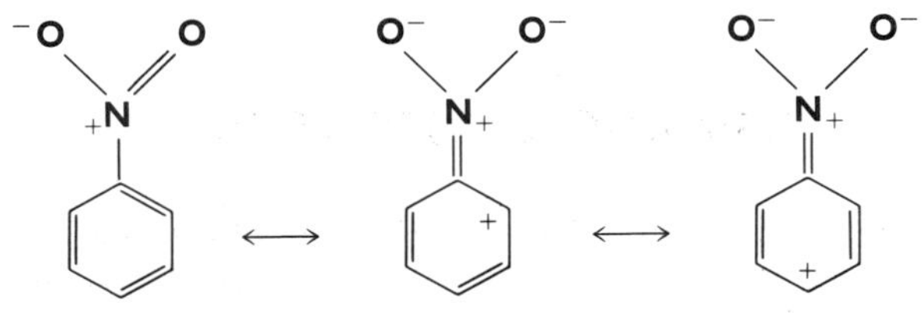

Both the *ortho* and *para* mononitro compounds give 2,4-dinitro-1-chlorobenzene under vigorous reaction conditions. The two most important factors are the temperature and the concentration of NO_2^+. At room temperature, very little dinitration occurs. To obtain the dinitro compound, the reaction is carried out at elevated temperature, and a larger amount of sulfuric acid is used. The excess sulfuric acid "ties up" the water formed in the reaction by converting it to H_3O^+, so that the series of acid-base equilibria that produce NO_2^+ are forced to the right.

Under either set of conditions, control over the extent of reaction is not total, and varying amounts of by-products are obtained. If the reactions are carried out with careful attention to conditions, the major product can be isolated and readily purified by crystallization. By-products can be identified by means of thin layer chromatography.

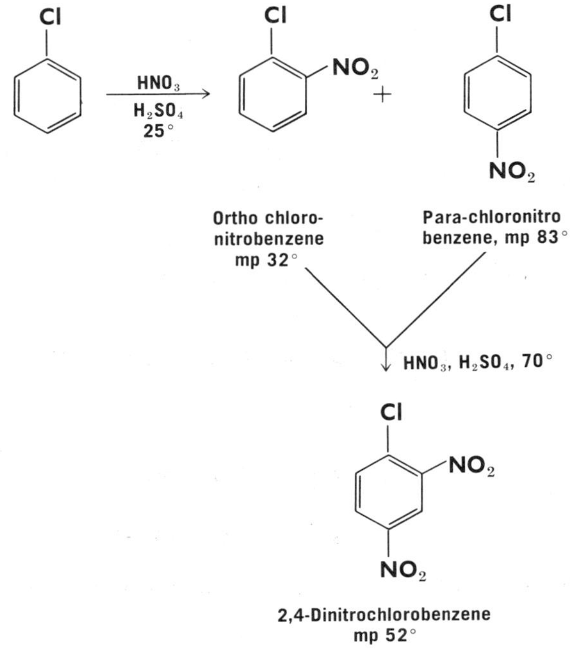

Ortho chloro-
nitrobenzene
mp 32°

Para-chloronitro
benzene, mp 83°

HNO_3, H_2SO_4, 70°

2,4-Dinitrochlorobenzene
mp 52°

EXPERIMENTS

The objectives of these experiments are to carry out the nitration of chlorobenzene under conditions leading to mono- or dinitro products, isolate the main product and identify by-products by TLC. Students should work in pairs, one doing the mononitration and the other the dinitration, so that results can be compared. Each student should label four small vials or 10×75 test tubes for TLC samples that will be obtained at various points in the procedures. For the mononitration, label the vials M-1, M-2, M-3 and M-4; for the dintration procedure, D-1, D-2, D-3 and D-4.

CAUTION: Concentrated nitric and sulfuric acids are used in these nitrations. Both of these acids can cause severe burns. Do not attempt to pour these acids while holding a graduated cylinder in your hand. If acid contacts the skin, immediately wash thoroughly with water.

Mononitration

In a 10 ml graduated cylinder obtain 8 ml of concentrated nitric acid. (Set the cylinder in a beaker before transferring the acid.) Pour the acid into a 125 ml Erlenmeyer flask and then obtain, in the same cylinder (without washing), 8 ml of concentrated sulfuric acid. Add the sulfuric acid slowly to the nitric acid and then swirl gently in cold water or an ice bath until the mixed acid has cooled to room temperature.

Obtain 3 ml of chlorobenzene. Using a pipet and bulb, add the chlorobenzene a few drops at a time to the acid, mixing thoroughly by swirling after each addition. Keep the mixture at room temperature by swirling in cold water or briefly in an ice bath, but do not cool below room temperature. After all the chlorobenzene has been added, swirl the mixture for another 5 minutes at room temperature and then cool it in an ice bath.

Add about 20 g of chopped ice to the flask and stir with scraping to break up the solid lumps of product. Allow any excess ice to melt and then collect the solid by suction filtration on a Hirsch funnel. Press the solid on the funnel and wash with a little water. Remove a small sample and dry on filter paper for a melting point, and place another small sample of the solid for TLC in vial M-1. Remove a drop of the oily layer in the filtrate and place it in vial M-2.

Recrystallize the rest of the solid in a test tube (18×150 or 25×100). Add 5 ml of methanol, heat to boiling in a steam or hot water bath and then add just enough methanol (1 to 2 ml) to dissolve all of the solid at the boiling point. Allow to cool, and then chill in an ice bath, collect the solid by suction filtration and spread the crystals on glassine paper to dry. *crystallized*

Add water to the methanol filtrate and transfer a small sample of the *yellowish* solid or oil (or mixture) which separates to a vial M-3. Place a small sample *when* of the recrystallized solid in vial M-4.

After carrying out the TLC examination, determine (at the same time) the melting points of the crude solid and the recrystallized product. Weigh the recrystallized product and calculate the percentage yield.

3 days clear cloudy pagean yellow   *after heating clear to yellow opaque yellow*

Dinitration

CAUTION: Avoid contact of the product with the skin; a burning sensation will result.

In a 10 ml cylinder, obtain 8 ml of concentrated nitric acid (set the cylinder in a beaker before transferring the acid). Pour the acid into a 125 ml Erlenmeyer flask. In a 50 or 100 ml cylinder, obtain 20 ml of conc. sulfuric acid; in a small test tube obtain 3 ml of chlorobenzene. Place the flask containing the nitric acid **in the hood** and slowly pour in the sulfuric acid. Swirl the flask to mix the acids, and then, without cooling, add the chlorobenzene in one portion. A vigorous exothermic reaction sets in, with evolution of brown fumes. Hold the flask by the neck, using a loop of towel if it is too warm to handle, and swirl steadily for 10 minutes to keep the layers well mixed. The heat of reaction will keep the mixture at the right temperature (about 60 to 70°).

After 10 minutes, the evolution of fumes should be nearly stopped. Cool the flask by swirling in an ice bath and add ice, a few pieces at a time at first, swirling to keep the mixture cold. Continue adding ice until the flask is about half full, and then stir with a glass rod until the product solidifies. When all the ice has melted, collect the product by suction filtration on a Hirsch funnel. Press the solid on the funnel and wash with a little water. Remove a small sample of the solid and dry on filter paper for a melting point. Place another small sample for TLC in vial D-1. If oily droplets are present in the aqueous filtrate, pour out the aqueous solution and scrape or rinse the oil with a few drops of acetone into vial D-2. If no oil is present, leave vial D-2 empty.

Transfer the rest of the solid to a test tube (18 × 150 or 25 × 100) for recrystallization. Add about 5 ml of methanol and warm until the solid melts and forms an oily lower layer. While the solution is warm, add just enough more methanol, stirring after each ml is added, to give a homogeneous solution (one layer). Since the product has a low melting point, it will separate as an oil on cooling. Rub the oil droplets gently on the wall of the tube while the solution is at room temperature until crystals begin to form. Then cool, with stirring, and keep the mixture in an ice bath for several minutes. Collect the solid on a Hirsch funnel, press on the filter and allow the crystals to dry on glassine paper.

Add water to the methanol filtrate and pipet a drop of the oil that separates into vial D-3. Place a small sample of the recrystallized solid in vial D-4.

After carrying out the TLC examination, determine (at the same time) the melting points of the crude solid and the recrystallized product. Weigh the recrystallized product and calculate the percentage yield.

TLC Examination

Review the TLC procedure in Chapter 7. The TLC analysis of the products is carried out with 2.5 × 7 cm strips of silica gel sheet. A suitable solvent system for the chloronitrobenzenes is hexane-chloroform (9:1).

Obtain three TLC strips and a flip-top TLC jar. Add enough solvent mixture to cover the bottom (about 3 to 4 ml) and cap tightly. Prepare four capillaries for spotting the plates as described on page 13 or page 64.

The four vials in each series, mono- or dinitration, correspond to the same fractions:

1. Crude solid from reaction.
2. Filtrate from crude solid.
3. Oil from filtrate of recrystallization.
4. Recrystallized product.

Samples M-4 and D-4 should be pure *para*-chloronitrobenzene and 2,4-dinitrochlorobenzene, respectively, and can be used as comparison samples for identification of by-products in the other samples in both series. In a fifth vial (which can be used by a pair of students for both series) obtain from the side shelf a sample of *ortho*-chloronitrobenzene; label this vial 5.

Dissolve the samples in M-1 to 4 or D-1 to 4 and 5 in about 0.5 ml of acetone and apply samples in three lanes on each TLC plate in the sequence given below. Use samples from another student for the series that you did not do. Be sure to keep the spots small and well separated. Prepare the first plate and place it in the jar to develop while you apply samples to the second plate.

Lane	Mononitration Series			Dinitration Series		
	Left	Center	Right	Left	Center	Right
	Samples			Samples		
Plate 1	M-1	M-2	M-4	D-1	D-3	D-4
Plate 2	M-2	D-4	M-3	D-2 (or D-1)	M-4	D-3
Plate 3	M-4	M-2	5	M-4	M-2	5

When the solvent has risen nearly to the top of the strip, remove the strip, place the next one in the jar and snap on the cap. Allow the developed strip to dry for 30 seconds and then place it in a jar containing a few crystals of iodine. When the spots are visible, remove the strip and record the position and size of the spots in each lane.

NUCLEOPHILIC AROMATIC SUBSTITUTION

A less common but nonetheless useful reaction of certain aromatic compounds is **nucleophilic** substitution. A leaving group such as halogen on a benzene ring, as in chlorobenzene, is normally inert to substitution by either an S_N1 or S_N2 process (Chapter 10). If strongly *electron-withdrawing* substituents are present in positions *ortho* and *para* to the leaving group, however, an anion formed by attack of a nucleophilic reagent is stabilized, and displacement can occur. This situation is found in 2,4-dinitrochlorobenzene, and the chlorine in this compound is readily displaced by good nucleophiles. The same effect of the electron-withdrawing nitro groups which makes dinitrochlorobenzene resistant to further *electrophilic* substitution works in exactly the opposite direction to permit *nucleophilic* substitution.

One of the important applications of nucleophilic aromatic substitution is the synthesis of 2,4-dinitrophenylhydrazine, which is a very useful reagent for preparing derivatives of carbonyl compounds. Hydrazine is a particularly reactive nucleophile, and the reaction with 2,4-dinitrochlorobenzene is rapid and exothermic.

2,4-Dinitrophenyl
hydrazine

EXPERIMENT

Place 1 g of recrystallized 2,4-dinitrochlorobenzene in a 50 ml Erlenmeyer flask and dissolve it in 10 ml of methanol. Obtain 1 ml of 85% hydrazine hydrate solution (0.52 g N_2H_4 per ml) and dilute the hydrazine solution with 5 ml of methanol. Add this to the dinitrochlorobenzene solution and record your observations. After 10 minutes, collect the product by suction filtration, wash with a little methanol and record the weight of the air-dried crystals.

the grignard reaction

The Grignard reaction is one of the most general and useful synthetic methods in organic chemistry. Addition of a Grignard reagent, RMgX, to an *aldehyde* or *ketone* leads to a magnesium alkoxide which on hydrolysis gives an alcohol.

$$\underset{R_2}{\overset{R_1}{\diagdown}}C{=}O + R_3MgX \longrightarrow R_1{-}\overset{\overset{\overset{-}{O} \; \overset{+}{MgX}}{|}}{\underset{\underset{R_2}{|}}{C}}{-}R_3 \xrightarrow{H_3O^+} R_1{-}\overset{\overset{OH}{|}}{\underset{\underset{R_2}{|}}{C}}{-}R_3 + Mg^{++} + X^-$$

With an *ester*, the initial addition product is unstable because it contains two negative atoms on the same carbon, and a molecule of magnesium alkoxide is eliminated. The resulting ketone then undergoes further reaction with another Grignard molecule to give the tertiary alkoxide.

$$\overset{\overset{O}{\|}}{R_1C}{-}OCH_3 + R_2MgX \longrightarrow R_1{-}\overset{\overset{\overset{-}{O} \; \overset{+}{MgX}}{|}}{\underset{\underset{R_2}{|}}{C}}{-}OCH_3 \longrightarrow R_1{-}\overset{\overset{O}{\|}}{C}{-}R_2 + CH_3\overset{- \; +}{O}MgX$$

$$R_1{-}\overset{\overset{O}{\|}}{C}{-}R_2 + R_2MgX \longrightarrow R_1{-}\overset{\overset{\overset{-}{O} \; \overset{+}{MgX}}{|}}{\underset{\underset{R_2}{|}}{C}}{-}R_2 \xrightarrow{H_3O^+} R_1{-}\overset{\overset{OH}{|}}{\underset{\underset{R_2}{|}}{C}}{-}R_2 + Mg^{++} + X^-$$

In the formation of the Grignard compound, reaction of the organic halide and magnesium occurs on the metallic surface and is formally an oxidation of the metal. The reaction is usually carried out in ether. This solvent acts as a Lewis base, forming a coordination complex with the organometallic compound and permitting it to diffuse away from the metal. Several minutes are usually required for the initial reaction to begin, but as soon as an active surface forms on the metal, heat is evolved and the ether begins to reflux. The reaction is controlled by the rate of addition of the halide.

A major pitfall in preparing a Grignard reagent is the presence of alcohol, water or any other "active hydrogen" compounds, which react immediately with the reagent. This side reaction

$$RMgX + H_2O \longrightarrow RH + MgXOH$$

coats the metal with MgXOH and prevents attack of halide. It is almost invariably the cause of a Grignard reaction which "can't be started."

EXPERIMENT

The objective of this experiment is to prepare a Grignard reagent from bromobenzene and carry out the reaction of the reagent with methyl benzoate to obtain a sample of triphenylcarbinol.

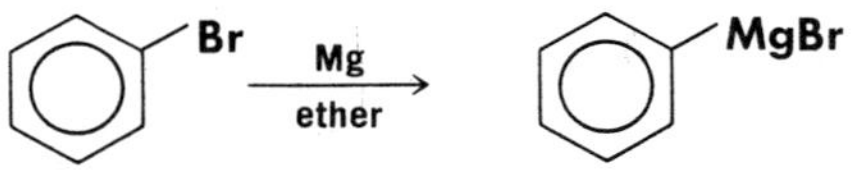

Bromobenzene

Phenyl
magnesium bromide

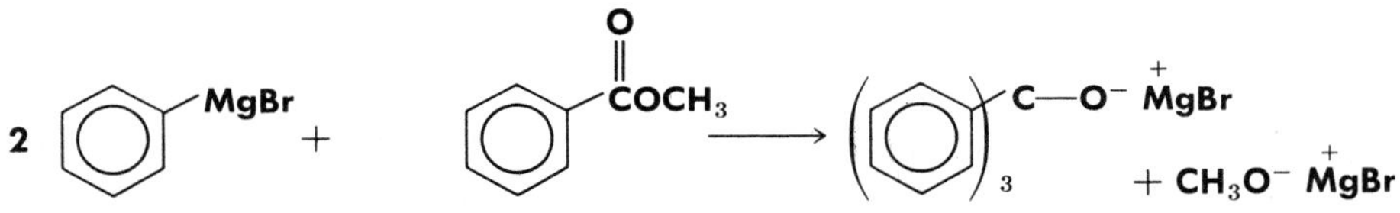

Methyl benzoate

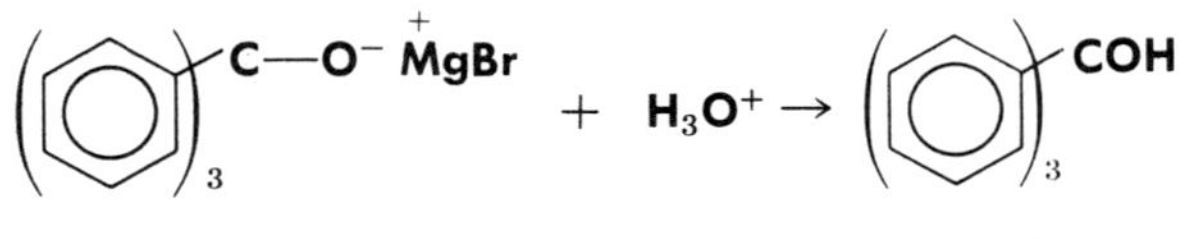

Triphenylcarbinol, mp 162°

A side reaction that usually occurs with aromatic Grignard reagents is a coupling reaction of the organomagnesium halide and unreacted aryl halide to give the hydrocarbon. Biphenyl is formed in the preparation of phenylmagnesium bromide. This byproduct is separated from the alcohol by crystallizing the product from a hydrocarbon solvent.

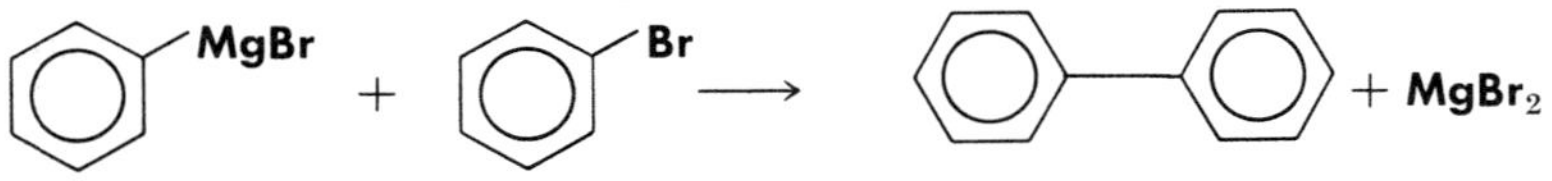

Biphenyl

Procedure

Clamp a three-neck, 500 ml round-bottom flask by the center neck to a ringstand, leaving room to slide an ice pan under the flask (do not support the flask on a ring). Place a reflux condenser in one of the side necks, a dropping funnel in the center neck and a stopper in the third neck. *All of the apparatus must be thoroughly dry,* with no traces of water or acetone. Fit a drying tube, containing calcium chloride supported on a cotton wad, in the top of the condenser.

When the apparatus is ready (stopcock on funnel closed; condenser tubing connected but water not running), obtain a pan of ice and 2.4 g of magnesium turnings. In *dry* graduated cylinders, measure out 50 ml of anhydrous ether and 9.5 ml of bromobenzene. Place the magnesium and 20 ml of ether in the flask. Draw up about 2 ml of the bromobenzene with a pipet and bulb and add this in a concentrated pool at the surface of the metal by inserting the pipet below the surface of the ether before squeezing the bulb. Stopper the flask and warm gently with the palm of your hand. If no visible sign of reaction (bubbling or cloudiness) is noted after 4 to 5 minutes, add a very small crystal of iodine. If the reaction still fails to start, add a further 2 ml portion of bromobenzene at the metal surface.

When the reaction begins, turn on a gentle flow of condenser water. Dilute the remaining bromobenzene with anhydrous ether, pour it into the funnel and add the remaining ether to the funnel. When refluxing ether begins to drip from the condenser, add the ether-bromobenzene mixture from the funnel at a rate just sufficient to maintain the refluxing. If the reaction becomes too vigorous, cool with the ice bath. As soon as all of the bromobenzene is added, close the stopcock, place a steam bath under the flask, and heat just enough to maintain gentle refluxing for another 10 minutes. Meanwhile, measure out 6.2 ml of methyl benzoate.

Remove the steam bath and cool the flask in ice until it is below room temperature. Pour the methyl benzoate into the dropping funnel, rinsing with a few milliliters of anhydrous ether, and add the ester dropwise to the Grignard solution; swirl the flask gently and cool in an ice bath if the ether begins to reflux. After the ester has been added, warm the reaction again and reflux 10 to 15 minutes. Meanwhile, obtain 60 ml of 4 N sulfuric acid and place 50 ml of it in a 250 ml Erlenmeyer flask.

Cool the reaction mixture again, remove the condenser and funnel and pour the contents of the flask into the aqueous acid. Rinse the three-neck flask with the remaining acid plus 20 ml of solvent ether and combine the two portions. The experiment can be interrupted at this point; cork the Erlenmeyer flask securely before storing.

Stir the mixture of ether and acid and pour into a separatory funnel; dissolve any solid remaining in the flask in a little more acid and ether and add this to the funnel. Drain off the aqueous layer, shake the ether layer with 10 ml of water, separate the layers cleanly, repeat the water washing and transfer the ether layer back to the (rinsed) Erlenmeyer flask. Add a teaspoonful of anhydrous magnesium sulfate, swirl and filter through cotton into a dry 250 ml Erlenmeyer flask.

To evaporate the ether without filling the laboratory with fumes, the vapors must be drawn into the aspirator. Place the flask (with boiling stone) on a steam bath and clamp a short length of glass tubing above the flask, with the end below the rim of the neck (see Fig. 1.2). Attach the glass tubing to the aspirator arm, turn on the aspirator and then begin to evaporate the ether.

When about one-half of the ether has been evaporated, add 25 ml of hexane and then resume evaporation until crystals begin to appear in the solution. Allow the flask to cool in air to room temperature, and then cool in an ice bath. Collect the product by suction filtration, determine the weight and melting point and calculate the yield. Submit the product to your instructor and place the mother liquor (filtrate from the crystallization) in the container provided.

Date ____________________

Name ____________________

Section ____________________

REPORT: CHAPTER 13

1. Description of Reactions

Give a brief account of your experience and observations, any problems with starting the Grignard reaction, appearance of reaction mixture at various stages, and isolation of product.

2. Data

Weight of vial + product:

Weight of vial:

Weight of product:

Mp of product:

3. Yield Calculations

Densities: bromobenzene 1.49; methyl benzoate, 1.09.

Moles of bromobenzene:

Gram-atoms of magnesium:

Moles of methyl benzoate:

Limiting reagent:

Moles product:

% Yield of product:

QUESTIONS

1. Write balanced equations with structural formulas for the reactions of phenylmagnesium bromide with:

    a. Acetone:

    b. Methanol:

    c. Benzaldehyde:

    d. Dimethyl carbonate ($CH_3O-\overset{\displaystyle O}{\overset{\|}{C}}-OCH_3$) (+ excess Grignard):

2. The methyl group of the methyl benzoate does not appear in the final product isolated. In what compound does the methyl group appear and *where* (what fraction or solution) is this compound at the end of the procedure?

3. Where is the biphenyl by-product at the end of the procedure?

4. After hydrolysis of the reaction mixture, the ether layer is separated and washed with water, and magnesium sulfate is then added. What is accomplished in each of these steps?

sulfanilamide

Sulfa drugs were among the first compounds available for treatment of bacterial infections, and they are still in wide clinical use for certain conditions, particularly urinary tract infections. Sulfa drugs are sulfonamides of p-aminobenzenesulfonic acid (sulfanilic acid). Most of the compounds now in use, such as sulfathiazole and sulfisoxazole, are sulfonyl derivatives of heterocyclic amines. The simplest member of the series, and the first to be used, is sulfanilamide, or p-aminobenzenesulfonamide.

Sulfanilamide Sulfathiazole Sulfisoxazole (Gantrisin)

The synthesis of sulfanilamide illustrates the use of a "blocking" or "protecting" group to provide for the presence of a reactive functional group in the final product of a several-step synthesis. The steps in the synthesis, starting with aniline, are shown in the formula scheme.

The NH_2 group of aniline is required in the final drug, but the amino group would be the point of reaction if chlorosulfonation were carried out directly with aniline. The $-NH_2$ group cannot be present during the introduction of the $-SO_2Cl$ group or while the $-SO_2Cl$ group is present. Therefore, the first step of the synthesis (already carried out in Chapter 3) is introduction of an acetyl group to block the reactivity of the $-NH_2$. The next step is chlorosulfonation, which is a typical electrophilic substitution reaction similar in mechanism and procedure to nitration (Chapter 12). The sulfonyl chloride is obtained directly by using excess chlorosulfonic acid. N-Acetylsulfanilyl chloride is the intermediate for preparing the whole family of N-substituted sulfa drugs. It is therefore commercially available and can be used as the starting point for the experiment in this chapter.

For the preparation of sulfanilamide, the sulfonyl chloride is treated with ammonia to give the sulfonamide. The final steps are then removal of the acetyl group, whose function has now been fulfilled, and neutralization to give the final product. In the hydrolysis step, both carboxamide ($-CONH-$) and sulfonamide ($-SO_2NH_2$) groups are present, but only the former reacts since the hydrolysis of sulfonamides requires drastic conditions. After acid hydrolysis, the sulfonamide is present as the soluble hydrochloride salt. Neutralization with sodium carbonate liberates the free amine.

In the last step, the solution must not be made strongly basic, as is usually done when converting an amine salt to the free amine, since an $-NH-$group in a sulfonamide is *acidic*. Because of the very strongly electron-withdrawing sulfonyl group, a sulfonamide dissociates in base to give an anion. Sulfanilamide is therefore an amphoteric compound, resembling p-aminobenzoic acid in its solubility properties. The antibacterial action of sulfanilamide is due to the fact that the compound interacts with an enzyme system which in bacterial metabolism converts *p*-aminobenzoic acid to a growth factor, folic acid. *p*-Aminobenzoic acid is a normal metabolite in the bacterial cell; sulfanilamide and other sulfa drugs, because of their structural similarity, are **antimetabolites**, and by blocking the enzyme system, prevent growth and multiplication of the bacteria.

| Cation in acid solution | Neutral | Anion in basic solution |

EXPERIMENT

The objective of this experiment is the preparation of sulfanilamide.

N-Acetylsulfanilyl Chloride (p-Acetamidobenzenesulfonyl Chloride)

Place 5.0 g of acetanilide in a *dry* 125 ml Erlenmeyer flask. To permit addition of the reagent in one portion without the reaction becoming too vigorous, the acetanilide is melted to reduce the surface area. Warm the bottom of the flask over a low flame until the solid melts, then tip the flask and rotate it while cooling to form a film spread over the bottom and the lower 1/4 inch of the flask. This step may be done in advance; stopper the flask securely until the reaction is completed.

Prepare a trap as illustrated in Figure 14.1 to prevent the escape of HCl fumes. Connect a length of rubber tubing from a one-hole rubber stopper in the Erlenmeyer flask to the stem of a glass funnel, which is clamped upside down with the rim just below the surface of the water. This is a simple way to handle water-soluble fumes; the system is sealed by water, but the water cannot back up into the reaction flask because of the large empty volume of the funnel.

In a *dry* graduated cylinder or test tube (depending on the dispensing

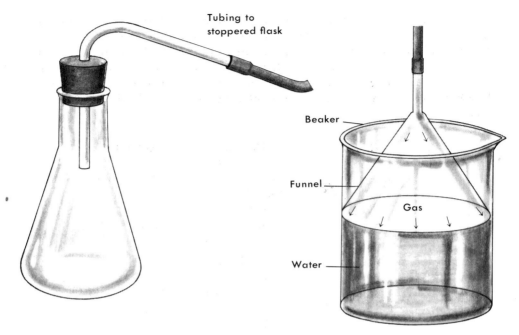

FIGURE 14.1 Trap for HCl gas.

arrangements), obtain 13 ml of chlorosulfonic acid, taking exactly the required amount. (**CAUTION:** This reagent is extremely hazardous and will cause severe burns. **DO NOT** pour this liquid into a sink.) Cool the acetanilide in an ice bath, add the chlorosulfonic acid in one portion and connect the flask to the trap. Remove the flask from the ice bath and return it only if the reaction becomes too vigorous. After 10 to 15 minutes, when most of the solid has dissolved, warm the flask on the steam bath; continue heating for about 10 minutes after the solid has disappeared. Cool the liquid (trap still connected), and then pour it onto 75 to 100 ml of ice. (Add water to the flask and transfer the residue to the main batch.) Stir and rub the solid sulfonyl chloride with a spatula until the lumps are broken up. Collect the product by suction filtration on a Büchner funnel and wash with a little water.

N-Acetylsulfanilamide (p-Acetamidobenzenesulfonamide)

Use the moist sulfonyl chloride prepared in the first step or weigh out 7.0 g of commercial N-acetylsulfanilyl chloride, as directed by your instructor. Place the sulfonyl chloride in a 125 ml Erlenmeyer flask (the same flask used in step 1 can be used again without washing). Add 20 ml of concentrated aqueous ammonia. Put fresh water in the beaker of the gas trap, and connect the trap again to avoid escape of ammonia fumes. Swirl the flask and warm on the steam bath for 5 to 10 minutes. Cool the mixture in ice and collect the sulfonamide by suction on a Büchner funnel. Wash the solid with water and press with a spatula.

Sulfanilamide

Transfer the cake of acetylsulfanilamide to the same Erlenmeyer flask and add 10 ml of water plus 5 ml of concentrated hydrochloric acid. Heat to gentle boiling over a low flame until all of the solid has dissolved, and continue heating for another 5 to 10 minutes. While waiting, prepare a fluted filter (p. 22). During the heating, add small amounts of water to replace any lost by evaporation. (This hydrolysis step can be carried out in a round-bottom flask with reflux condenser if desired, but the latter set-up is not needed.) Cool the solution, add 5 ml of water and a spatula of decolorizing carbon. Warm and then filter the solution through a fluted filter into another 125 ml Erlenmeyer flask. Cool again and add solid sodium bicarbonate in small portions, stirring after each addition, until the solution is neutral to test paper. Collect the solid by suction filtration. Save a small sample of this material for melting point determination, and allow it to air-dry.

Recrystallize the remaining product from water. Place the moist product in a 125 ml Erlenmeyer flask, add 30 ml of water and heat it to boiling. Add more water as needed to dissolve all of the solid, filter the solution rapidly through a small plug of cotton if it is not completely free of floaters, and chill. Collect the crystals and air dry them until the next laboratory period. Weigh the product, determine the yield and determine the melting point of the crude and recrystallized product.

REPORT: CHAPTER 14

1. Description of reactions

Note the course and behavior of reactions, changes in appearance, any difficulties encountered, any mechanical losses, and so forth, in any steps.

2. Product

Appearance:

Weight of paper + product:

Weight of paper:

Weight of product:

Melting points
  Crude product:

  Recrystallized:

3. Yield calculations

Moles of acetanilide (or N-acetylsulfanilyl chloride):

Moles of sulfanilamide:

% Yield:

QUESTIONS

1. What product would be obtained if aniline, rather than acetanilide, were treated with chlorosulfonic acid?

2. What would be observed if excess NaOH solution, rather than NaHCO$_3$, were added to neutralize the HCl in the last step?

chemistry of milk

Milk supplies the complete nutrition of young mammals, and the lactating mammary gland has been ranked second only to the photosynthesizing plant cell in the chain for sustaining life. The basic elements of all foods are proteins, carbohydrates and fats. These compounds are present in milk in amounts balanced for the needs of each species. The proportions vary rather widely, as seen in Table 15.1. Human milk is rich in carbohydrate but relatively low in protein, and carbohydrates must be added to cow's milk for feeding human infants. Significant differences in the concentration of total solids occur between various breeds of dairy cattle. The quality of milk is usually judged by the fat content, and a minimum fat content is specified by government regulations; a minimum of 3.25% is required in many localities.

The separation of milk into its components for the production of butter, cheese and whey has been carried out for many centuries. Most of the fat can be obtained simply by skimming the layer which rises after standing, or by centrifuging. To prevent cream separation, most whole milk is homogenized by pumping it through a small orifice which breaks up fat globules to such a small size that a stable colloidal suspension results. To make cheese, the major protein, **casein**, is precipitated, together with the fat if it has not been removed. The carbohydrate **lactose** remains in the whey, or filtrate, together with a mixture of salts, which includes about 12 metallic cations plus chloride, phosphate and sulfate. A substantial amount of the phosphate in milk is precipitated with the casein in the form of phosphate esters of hydroxyl groups in the protein.

TABLE 15.1 COMPOSITION OF MILK (BY WEIGHT)

	Water	*Protein*	*Fat*	*Carbohydrate*	*Mineral*
Human	87.4	1.4	4.0	7.0	0.2
Sheep	82.6	5.5	6.5	4.5	0.9
Cow (average)	87.1	3.4	3.9	4.9	0.7
Jersey	85.0	4.0	5.4	4.8	0.8
Shorthorn	87.3	3.3	3.9	4.8	0.7
Holstein	88.0	3.0	3.4	4.9	0.7

Isolation of the main components of homogenized milk in a laboratory procedure depends on the totally different solubility behavior of the three substances. The polypeptide chain in casein contains more acidic than basic groups, and casein is soluble as a polymeric anion at the pH of whole milk (about 6.0). The protein is very much less soluble at the isoelectric pH, 4.7, where the anionic and cationic charges are balanced, and addition of a small amount of acetic acid causes the casein to precipitate. The colloidal particles of butterfat, which are kept in suspension by the presence of protein, agglomerate and separate with the casein. The steps in the separation process are outlined in the following diagram.

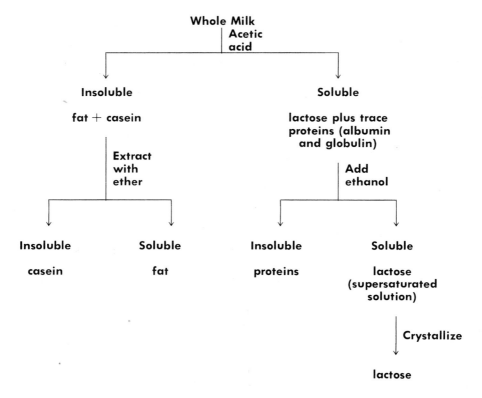

After precipitation, the mixture is strained and the insoluble curd is extracted with acetone and ether. In this step, the acetone acts as a co-solvent for both water and ether so that the fat globules dispersed in the curd can be dissolved in the ether. The extract is then evaporated completely to remove the acetone, and the fat is isolated.

The butterfat obtained by this procedure is "clarified butter," containing about 98% triglycerides plus small amounts of other lipids (Chapter 18), including the fat-soluble vitamins A and D. The distinctive flavor of butter is due to traces of the diketone $CH_3COCOCH_3$ (biacetyl) and the α-hydroxyketone $CH_3CHOHCOCH_3$. In contrast, creamery butter contains about 14% of water emulsified in the fat plus 1 to 2% of protein.

The recovery of butterfat by this solvent extraction procedure is not complete, since some fat remains trapped in the casein. Quantitative analysis for butterfat is an important determination, because milk is graded and purchased on the basis of fat content. This analysis is carried out routinely by the Babcock method. A sample of milk is treated with a large amount of concentrated sulfuric acid to dissolve all of the protein and other constituents except fat, and the volume of the fat layer is measured in a special flask with a calibrated neck.

The carbohydrate in milk is almost exclusively one sugar — the disaccharide lactose. The aqueous whey contains a small amount of soluble proteins, lactalbumin and globulins, and these must be removed before the lactose is isolated. In commercial practice, the proteins are separated by neutralizing the solution with lime and heating until a compact mass of protein coagulates. The lactose is then isolated by evaporation of most of the water. In the laboratory procedure, the protein is precipitated by diluting the whey with a large volume of ethyl alcohol. After removal of the protein, the lactose crystallizes directly in nearly pure form from the alcohol solution.

EXPERIMENTS

The objective of the experiments in this chapter is to separate and isolate the three major components of milk. The properties of these substances will be examined in experiments in Chapters 16, 17 and 18. One of these experiments is the isolation of mucic acid by oxidation of whey. The chemistry underlying this oxidation and the procedure are discussed in Chapter 16 (pp. 151 and 155). The reaction can be carried out while the other milk constituents are being isolated, and it is advantageous to proceed to this part of Chapter 16 during the present experiment.

ISOLATION PROCEDURE

Place 50 ml of homogenized fresh milk in a 125 ml Erlenmeyer flask and warm in a hot water bath to 40°. Using a pipet and bulb, add about 1 ml of 50% acetic acid to the milk, swirling after every few drops, until a curdy precipitate forms. Allow the mixture to cool while the curd coagulates. Fold a 1 square foot piece of cheese cloth into a 4 to 6″ square pad of several thicknesses, place the cloth over a 250 ml beaker and strain the coagulated milk through the cloth. Squeeze the gummy mass firmly. Transfer the filtrate (whey) to a graduated cylinder and record the volume. Use part of this solution for isolation of lactose, and part of it for oxidation, as described in Chapter 16.

Transfer the solid in the cheese cloth back to the Erlenmeyer flask, scraping as much material as possible from the cloth, for isolation of butterfat and casein.

1. Isolation of Lactose

Place 10 ml of the cloudy whey solution in a 125 ml Erlenmeyer flask. Heat the whey to boiling while swirling, then turn off the burner and add 100 ml of absolute (anhydrous) ethanol (95% ethanol can be used if absolute alcohol is not available; the latter is preferable). Filter by gravity through paper into a dry 125 ml Erlenmeyer flask. If the initial filtrate is turbid, pour it back into the funnel; the filtrate should be free of all suspended solid. The filtration takes some time to complete since the precipitated protein gradually clogs the filter. Portions of the filtrate can be removed during the filtration, if desired, for use in experiments in Chapter 16. When the filtration is complete, cork the flask and allow it to stand undisturbed. Record the appearance of the crystals that appear and then scrape the walls and bottom of the flask to initiate additional crystallization sites. Set the flask aside for at least 24 hours, and then scrape the crystals loose and collect by suction filtration on a Hirsch funnel. Record the weight of the lactose, and from the volume of the original whey, calculate the weight that would have been obtained from the entire sample. Determine the melting point of the crystals.

2. Isolation of Butterfat

To the milk solid in the Erlenmeyer flask add 10 ml of acetone and 10 ml of ether (**CAUTION! NO FLAMES**). Stir the mixture thoroughly, smoothing out lumps with a spatula, and pour off the solvent layer into a 100 ml round-bottom flask (to permit evaporation at reduced pressure, do not use an Erlenmeyer flask). Add 15 ml of ether to the solid and again stir and decant into the same flask. Repeat with an additional 10 ml of ether.

Add a boiling stone to the ether solution and evaporate the ether on a steam or hot water bath, either in the hood or by using a tube attached to the aspirator held over the neck of the flask to sweep ether vapor into the aspirator (Fig. 1.2, page 4). When the volume has been reduced to about 10 ml, attach the flask to the aspirator with a rubber stopper and evaporate under reduced pressure until the residue is a clear yellow oil plus a cloudy water layer. While the last traces of acetone are evaporating, press a small wad of cotton into the top of the stem of a funnel and cover it with a thin layer of anhydrous sodium sulfate. Place a weighed 25 × 100 ml test tube under the funnel.

Add 5 ml of ether to the oily residue and transfer the upper ether layer with a pipet and bulb to the funnel. (When some of the lower layer is drawn up into the pipet, let this drain back into the flask first. A separatory funnel can be used, but it is not necessary, and a much larger amount of solvent is needed for rinsing.) Rinse the flask and the sodium sulfate with a few 1 ml portions of ether. Evaporate the dried ether solution in the test tube to a thick oil and remove the last traces of solvent with aspirator vacuum. Weigh the residual butter and record the yield. If necessary, chill the oil in cold water to congeal the butter.

3. Casein

To the solid remaining in the Erlenmeyer flask after ether extraction, add 15 to 20 ml of acetone, stir and smooth lumps and decant the acetone into a Büchner funnel set up for suction filtration. Stir the solid with another 20 ml of acetone, rub well and then collect the solid on the funnel. Allow the solid to air dry and record the weight.

carbohydrates

Carbohydrates, formed by photosynthesis in green plants, make up the bulk of plant tissue and are the mainstay of all animal diets. The basic unit of carbohydrates is a **pentose** or **hexose** sugar. **Disaccharides** contain two of these units, and **polysaccharides** contain a large number of units. The most important sugar is **D-glucose**, which is present in every living cell and is the repeating unit in the polysaccharides starch and cellulose. Glucose is an aldohexose and exists as a cyclic *hemiacetal* in equilibrium with a small amount of the aldehyde form. Since the aldehyde carbon becomes an asymmetric center in the cyclic form, there are two stereoisomeric hemi-acetals, called α and β.

α-D-glucose   Aldehyde form   β-D-Glucose

DI- AND POLYSACCHARIDES

The sugar units in di- and polysaccharides are linked by *acetal* linkages, called **glycoside bonds,** formed between the hemiacetal carbon in one sugar and a hydroxyl group in the next unit. In starch, the acetal link between glucose units has the α-configuration; in cellulose the acetal group is β. The difference in configuration of the two polymers causes a marked contrast in properties. The polymer chain in starch is branched and has a coiled structure; starch is fairly soluble in water and forms a deep-blue complex with iodine. Cellulose has a much higher molecular weight; the chain is completely unbranched and forms strong fibers.

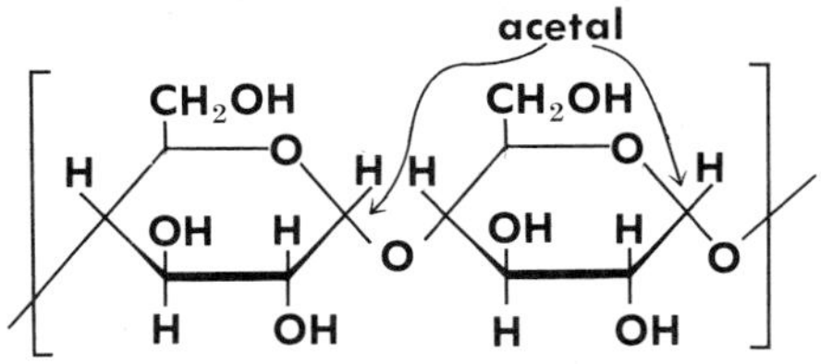

Repeating unit of starch

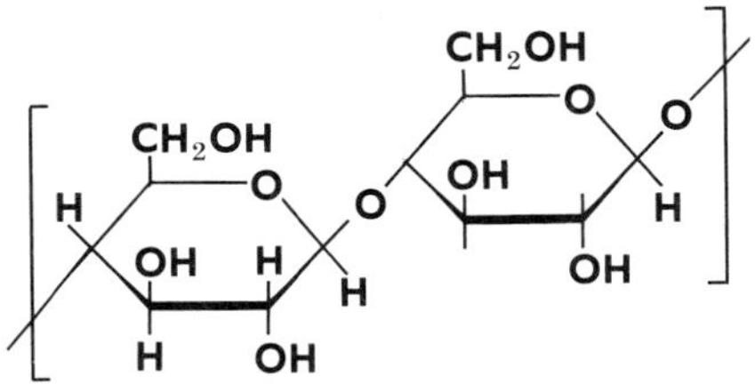

Repeating unit of cellulose

The glycoside linkages in disaccharides and polysaccharides are hydrolyzed in acid solution to give the constituent sugars. Hydrolysis can also be accomplished very rapidly by enzymes, which are proteins with highly specific catalytic activity for specific bonds. Amylase enzymes in saliva and gastric juice bring about rapid hydrolysis of the α-glycoside bonds in starch, but do not affect the β-glycoside groups in cellulose.

The important disaccharides **lactose** from milk, and **sucrose**, or table sugar, contain glucose as one unit and galactose and fructose, respectively, as the second sugar. In lactose, the glucose unit is present as the hemiacetal, in equilibrium with the aldehyde form; lactose therefore exists in α and β forms. In sucrose, both the glucose and fructose units are linked through the carbonyl carbons in a "double" acetal group which is very susceptible to acid hydrolysis. Sucrose is also hydrolyzed by the action of invertase, an enzyme found in yeast and secreted by honey bees; honey contains a higher proportion of fructose than any other sweetening agent.

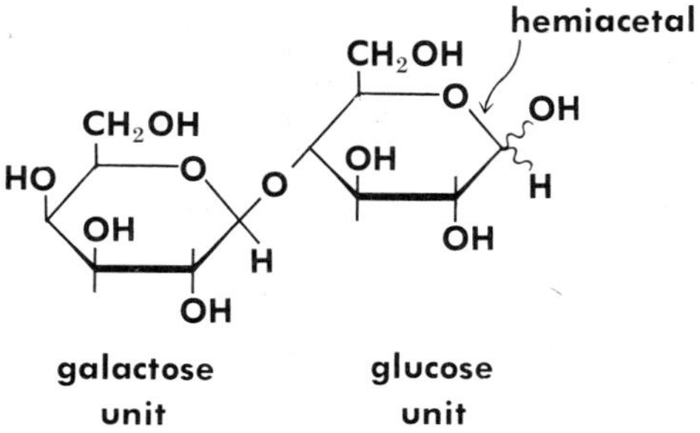

galactose unit — glucose unit
Lactose

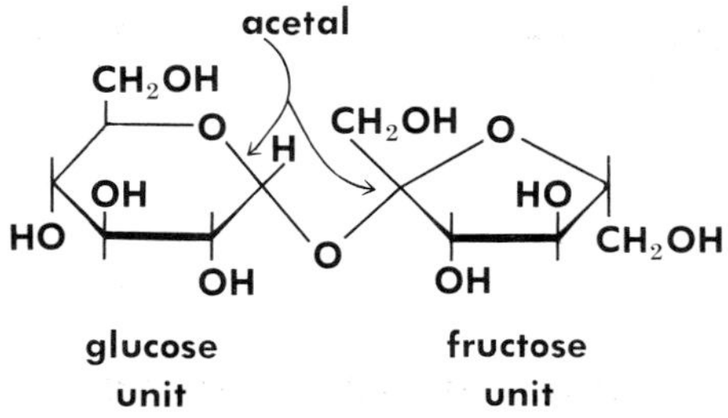

glucose unit — fructose unit
Sucrose

REDUCING PROPERTIES OF SUGARS

Carbohydrates, such as glucose and lactose, that have hemiacetal groups in equilibrium with the aldehyde form are **reducing sugars**. The aldehyde is oxidized by Cu^{++} or Ag^+ ion, and the formation of the reduction products red cuprous oxide (Cu_2O) or a silver mirror (Ag^0) is a sensitive test for a sugar that contains a hemiacetal $\leftrightarrows$ aldehyde group. A useful reagent for observing reducing properties is Benedict's solution, which contains Cu^{++} ion in base plus citrate ion to prevent precipitation of cupric hydroxide. All hexose sugars and lactose, but not sucrose,

$$R-\overset{\overset{\text{O}}{\|}}{C}-H + 2\ Cu^{++} + 3\ OH^- \longrightarrow R-\overset{\overset{\text{O}}{\|}}{C}-O^- + Cu_2O + 2\ H_2O$$

Blue
solution

Red
precipitate

reduce Benedict's reagent. Quantitative methods based on the reduction of Cu^{++} ion are used routinely in clinical laboratories to determine glucose in body fluids.

IDENTIFICATION OF GALACTOSE

Reducing power is not specific for a given sugar, and other methods are needed to identify individual compounds. An example is **galactose,** which occurs in the urine of infants with a genetic disease in which enzymes needed for metabolism of this sugar are absent. Another situation in which the identification of galactose is required is in detection of milk solids in meat products; the use of dried milk as a filler in meat and ice cream is prohibited in many localities to avoid adulteration and violation of dietary restrictions. Since lactose in milk is the only edible form in which galactose occurs, the detection of galactose is proof of the presence of substances derived from milk.

Identification of galactose, or lactose, is based on the oxidation of galactose by dilute nitric acid to the dicarboxylic acid, called **mucic acid.** This reaction was first observed by Scheele in 1780, during the early dawn of organic chemistry, and this type of oxidation played an important role in determining the structures of the hexoses. The diacids obtained from glucose and another aldohexose, mannose, are asymmetric compounds and are optically active. The acid from galactose, however, is a *meso* compound, with a plane of symmetry, and has zero optical rotation. The results of these oxidations provided the key evidence for the configuration of the hydroxyl groups in glucose, galactose and, indirectly, other sugars. The diacid obtained from nitric acid oxidation of glucose is quite soluble in water. However, mucic acid from galactose, with a more symmetrical molecule, forms crystals with a very high melting point which dissolve only to the extent of

0.3% in water. The isolation of mucic acid from aqueous solution is therefore very simple, even when other similar compounds are present, and mucic acid can be obtained directly by oxidation of the whey from milk.

| Glucose | Saccharic acid; $\alpha]_D +20°$; mp 120° | Galactose | Mucic acid $\alpha]_D 0.0°$; mp 260° |

OSAZONES

Since sugars are extremely soluble in water, it is often difficult to isolate a sugar directly. A general reagent for obtaining water-insoluble derivatives is phenylhydrazine. Hydrazones are formed by condensation of the aldehyde group with one mole of phenylhydrazine. With excess reagent, oxidation of a hydroxyl group and further condensation occur to give an **osazone**, which is a derivative of the 2-keto aldehyde. Osazones of some hexose sugars crystallize directly from hot water as they are formed; the time required for separation and the characteristic appearance can be used for tentative identification.

| Hydrazone | | Osazone |

FRUCTOSE

Fructose, one of the two sugar units in sucrose, is a 2-ketohexose. Fructose and glucose are closely related, and in the metabolism of carbohydrates a phosphate ester of glucose is converted to fructose phosphate. The interconversion of glucose and fructose occurs by *enolization* of the carbonyl groups. Both sugars can form the same enediol, and in the presence of base, either sugar leads to an equilibrium mixture. Fructose gives the same osazone as glucose, and because of the rapid interconversion to glucose in base, fructose reduces Benedict's solution as rapidly as an aldohexose.

Fructose ⇌ Enediol ⇌ Glucose

In *acid* solution, equilibria between glucose and fructose are established much more slowly; thus, glucose can be isolated from the acid hydrolysis of starch without conversion to fructose. In very strong acid, all sugars undergo a complex series of reactions leading eventually to an aromatic compound, 5-hydroxymethylfurfural. This process involves several enolization and dehydration steps, and takes place much more rapidly with fructose than with aldohexoses.

The Seliwanoff test for fructose depends on the formation of hydroxymethylfurfural in 4 N hydrochloric acid in the presence of resorcinol, which condenses to give a red dye. Sucrose is rapidly hydrolyzed under these conditions, and both sucrose and fructose give a cherry-red color. Glucose and other aldohexoses give a faint pink color because of their slow enolization. Pentose sugars such as arabinose are converted to furfural in this test, and a grey-green color results from condensation with resorcinol.

EXPERIMENTS

Several experiments are described in this chapter to permit observation of various aspects of carbohydrate chemistry and provide experience with the important reactions of sugars. Each student will probably not carry out all of the experiments, but experiments 1, 2 and 5 can easily be completed in one laboratory period. Pairs of students may want to divide up the experiments and share results; 1 and 2 must be carried out as a unit.

Most of the procedures involve simple test tube reactions in a hot water bath. A number of solutions are required; some are used in several experiments and it is important to plan the sequence of work to be done to minimize the number of trips to the side-shelf. Several of the procedures can be done concurrently, and good organization and careful labeling of test tubes is essential.

1. Hydrolysis of Starch and Sucrose

In this experiment, the objective is to compare the effectiveness of acid and of the amylase enzyme from saliva in catalyzing the hydrolysis of starch. Completion of the reaction is observed by disappearance of starch as indicated by the blue iodine complex.

a. Acid Hydrolysis. In an 18×150 mm test tube place 10 ml of 1% starch solution (prepared by adding a paste of 1 g of soluble starch in a few milliliters of cold water to 100 ml of boiling water). Add 1 ml of 2 N HCl (or 2 ml of 1 N acid) to the test tube. Line up six 10×75 mm or 13×100 mm test tubes, and with a pipet and bulb, transfer 1 ml of the acidified starch solution to the first small test tube. To this sample add one drop of 0.002 M iodine solution (contains 0.1 g of I_2 and 0.3 g of KI in 200 ml of water). Note the color of this solution at "time zero."

Place the test tube containing the starch in a boiling water bath, and at 3 minute intervals withdraw a 1 ml sample with a pipet and add it to one of the small test tubes. Cool the sample to room temperature and add 1 drop of iodine solution. Continue heating and sampling for 15 minutes or until a test sample shows negligible blue color. Estimate and record the time when the starch has been completely hydrolyzed.

b. Enzymatic Hydrolysis. Place another 10 ml sample of 1% starch solution in a test tube. Rinse out the small test tubes, line them up and add one drop of iodine solution to each. To obtain a sample of enzyme, rinse your mouth with water and stimulate the flow of saliva by chewing briefly on a small piece of clean plastic sheet or tubing. Place a sample of saliva in a small beaker and then, noting the time, add 5 drops of saliva to the starch solution and mix by shaking. At 1 minute intervals pipet 1 ml samples into the tubes containing iodine solution and record the time at which the blue starch-iodine color is negligible. Save the remaining solution of hydrolyzed starch for the next experiment.

c. Hydrolysis of Sucrose. In three 18×100 mm test tubes labeled A, E and B, place 5 ml samples of a 1% solution of sucrose. To tube A add 1 ml of 1 N HCl. Heat the solution for 2 minutes in the hot water bath, cool, and neutralize the acid with 1 ml of 1 N NaOH. To tube E add 5 drops of saliva and allow the tube to stand 5 to 10 minutes. Tube B serves as a "blank," with no additions being made. Test the solutions for evidence of hydrolysis in the next experiment, and record the results.

2. Reducing Properties of Sugars

The objective of this experiment is to observe the Benedict test for reducing sugars, and to use this test to determine the results of the sucrose hydrolysis in Part 1.

In three 18×150 mm test tubes labeled 1, 2 and 3 place the following:

Tube 1: 5 ml of 1% glucose solution.
Tube 2: 5 ml of the 1% starch solution used in Part 1b.
Tube 3: The remaining starch solution from the enzymatic hydrolysis in Part 1b.

To each of the tubes 1, 2 and 3, and tubes A, E and B from Part 1c, add 5 ml of Benedict's solution (contains 17.3 g $CuSO_4 \cdot 5 H_2O$, 100 g of Na_2CO_3 and 175 g of sodium citrate per liter). Mix the contents of the tubes by shaking or stirring (rinse the rod each time!). Note the time and place the six tubes at the same time in a boiling water bath. After 3 minutes remove the tubes as close to the same time as possible and place them in cold water. Record the appearance of the tubes and interpret your results.

3. Mucic Acid from Whey

This experiment demonstrates the method used for the identification of lactose or galactose in a dilute mixture, such as whey in milk, galactosemic urine, or a food extract. The procedure given starts with whey in milk; alternatively, a solution of 50 mg of lactose in 2 ml of water can be used.

In a 25×100 mm test tube, place 2 ml of the whey obtained from milk after precipitation of casein and fat. Add 1 ml of concentrated nitric acid and place the tube in a steam bath or boiling water. Clamp a piece of glass tubing connected to the aspirator hose above the test tube as shown in Figure 1.2 (p. 4) to sweep NO_2 fumes into the aspirator. After 1 hour, or after the volume of the solution has decreased to about 1/2 ml, add 1 ml of water and then cautiously evaporate almost to dryness with a very low flame, taking care not to char the residue on the wall of the tube. Cool the syrup, add a few drops of water and scratch until crystals begin to form. Add a little more water, stopper the tube and allow it to stand as long as time permits. Collect the crystals on a Hirsch funnel, rinse with a little water and then with alcohol, dry in air and record the weight.

4. Osazones

The objective of this experiment is to observe the classical osazone test for sugars, and the relative rates of formation of the same osazone from two different sugars.

In labeled 18 × 150 mm test tubes place 0.1 g samples of glucose, fructose, lactose and sucrose, and dissolve in 1 ml of water. (If 10% solutions of the sugars are made up, use 1 ml of solution.) To each tube add 5 ml of phenylhydrazine reagent (prepared from 4 g of $C_6H_5NHNH_2 \cdot HCl$, and 6 g of $CH_3CO_2Na \cdot 3 H_2O$ in 40 ml of water).

Place the tubes at the same time in a boiling water bath and watch for reactions, noting the time when solid separates in any of the tubes. After 30 minutes, cool the solutions and record the appearance.

If time permits, collect the products by suction filtration on a Hirsch funnel, wash with a little cold water and allow the crystals to dry. Check the identity of any crystals from different tubes that appear to be the same by melting point and mixture melting point.

5. Fructose

This experiment demonstrates the relationship between sucrose, fructose and glucose in acid and alkaline solution. The objective is to observe the formation of fructose by two paths, using the Seliwanoff test for identification.

In labeled 18 × 150 mm test tubes, place the following solutions:

Tube 1: 1 ml of 1% fructose.
Tube 2: 1 ml of 1% sucrose.
Tube 3: 1 ml of 1% arabinose.
Tubes 4 and 5: 1 ml of 1% glucose.

To tube 5 add 0.5 ml of 1 N NaOH solution. After this solution has stood 10 minutes at room temperature, add 5 ml of Seliwanoff reagent (prepared by dissolving 50 mg of resorcinol in 33 ml of concentrated HCl and diluting to 100 ml with water). Add 5 ml of reagent to the other four tubes and then place all five tubes, at the same time, in a boiling water bath. Observe the tubes for several minutes and record the appearance after 1, 2, 5, 10 and 15 minutes.

Date ______________________________

Name ______________________________

Section ______________________________

REPORT: CHAPTER 16

1. Hydrolysis of Starch and Sucrose

a. Acid. Time for complete hydrolysis: ______

b. Enzymatic Amylase. Time for complete hydrolysis: ______

c. Sucrose Hydrolysis. Reducing properties (from Part 2) present or absent?

Tube A: ______

Tube E: ______

Tube B: ______

Does saliva contain invertase enzyme? ______

2. Reducing Properties

Tube	Substance Added	Appearance of Tube
1.		
2.		
3.		
A.		
E.		
B.		

Which of the six tubes appeared to contain the largest amount of Cu_2O?

What sugar or sugars are responsible for the reducing properties in tubes showing positive test? What substances are present in the tubes that do not show reducing properties?

3. Mucic Acid from Whey

Weight of product + paper:

Weight of paper:

Weight of product:

Molecular weight of mucic acid:

Moles mucic acid:

Molecular weight of lactose:

Percentage yield from whey, assuming 4.8% lactose:

4. Osazones

Appearance of tubes (time of separation of solid from solution):

1. Glucose:

2. Fructose:

3. Lactose:

4. Sucrose:

Melting point data:

5. Fructose

Appearance of tubes:

Tube	1 min	2 min	5 min	10 min	15 min
1					
2					
3					
4					
5					

Interpretation: Account for the results observed in tubes 2, 4 and 5 in terms of the scheme on page 153. Note any observations that do not seem to fall in line with the expected behavior.

QUESTIONS

1a. Saliva contains 99.5% water. Assume that salivary amylase accounts for 5% of the solid material present, and that the molecular weight of the enzyme is 10,000. Calculate the number of *moles* of enzyme that were used in Part 1b (5 drops $\cong$ 0.5 g of saliva).

b. The hydrolysis of starch to glucose can be represented by the equation:

$$(C_6H_{10}O_5)_n + nH_2O \longrightarrow nC_6H_{12}O_6$$

Calculate the number of *moles* of glucose formed from 0.1 g starch, assuming complete reaction.

c. How many molecules of glucose are liberated per molecule of enzyme?

2. In the acid hydrolysis of a disaccharide, where does the proton from the acid become attached to catalyze the reaction? Draw a structural formula and show precisely how the hydrolysis occurs.

3. Calculate the molar ratio of phenylhydrazine to hexose used in the osazone formation, based on the composition of the reagent and the amount of sugar and reagent that were used.

4. Honey contains sucrose, fructose and glucose. The Benedict test can be used for the quantitative estimation of reducing sugars, and the Seliwanoff test can be used for the quantitative determination of fructose and sucrose in the presence of glucose. Outline a strategy by which the three sugars in honey could be determined using these two methods.

proteins and amino acids

Proteins are biopolymers with chains of α-amino acid units linked together by amide or **peptide** bonds. In contrast to carbohydrate polymers, with only one monomer unit, proteins contain 20 different amino acids.

Peptide bonds

The polypeptide chains in proteins vary in length from about 50 amino acids to over 10,000.

Proteins have a number of functions or organisms. Insoluble **fibrous** proteins provide the elastic fibers in muscle and connective tissue, and the skin, hair, nails and horns of various animal species. **Globular** proteins, soluble in water, include serum proteins, albumins and the enzymes which are catalysts for the chemical reactions in metabolism, as observed in the action of salivary amylase in Chapter 16.

The properties of proteins depend on the molecular size, the nature of the R groups in the amino acids, and the sequence of the amino acids. Certain proteins, for example silk, have characteristic repeating sequences of a few amino acids which impart a particular structure. All proteins contain amino acids with basic groups, such as $-NH_2$ and $-NHC(=NH)NH_2$ (guanidine), and with acidic $-CO_2H$ groups. Other amino acids contain $-OH$ and $-SH$ groups and phenolic rings. Casein, the major protein of milk, is a phosphoprotein, with phosphate ester groups attached to the $-OH$ of an amino acid.

Proteins usually have minimum solubility at the isoelectric pH, with balanced anionic and cationic charges, and are more soluble at higher pH, with excess negative charges, or lower pH, with excess positive charge. Casein is an exception: because of the acidic phosphate groups, casein is soluble in basic solution but is precipitated by acid. Most globular proteins, such as albumin, are denatured and rendered insoluble simply by heating or by addition of organic solvents or heavy metal ions. Rapid agitation can also cause denaturation, as when egg whites are beaten to a meringue.

Proteins undergo reactions that are characteristic of the component amino acid side chains. The yellowing of skin by nitric acid, for example, is caused by a reaction of phenolic rings in the amino acid tyrosine, present in skin protein. The amino acid tryptophane, containing an indole ring, is very sensitive to acid and decomposes during acid hydrolysis of a protein, with formation of a deeply colored pigment. Proteins containing cysteine, with an SH group, and the corresponding disulfide cystine, lose sulfide ion on heating in base.

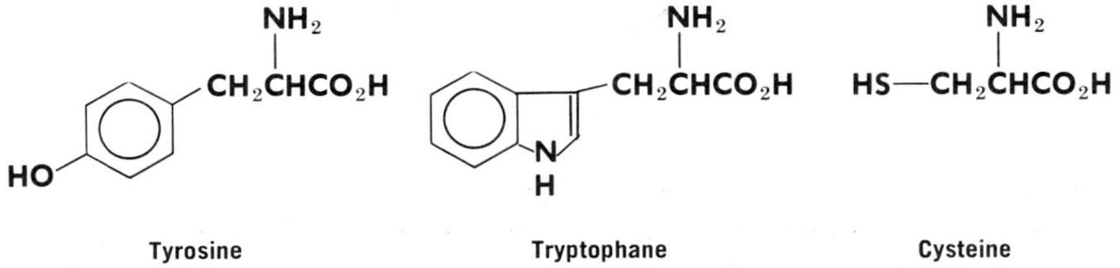

Tyrosine   Tryptophane   Cysteine

A general reaction which is useful as a qualitative test for proteins is the formation of a violet-colored complex with copper ion in alkaline solution. This test, called the **biuret** reaction, depends on the presence of two peptide bonds. The polypeptide chain can coordinate the metal ion at several points; one possible arrangement is that shown below:

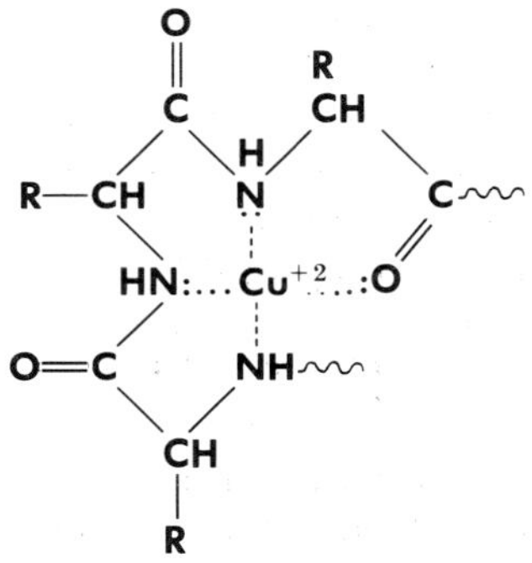

Proteins are broken down into the component amino acid units by hydrolysis, usually in hot 15 to 20% hydrochloric acid. The amino acids can then be separated and analyzed by chromatography. Qualitative examination of the amino acid mixture can be carried out by partition chromatography on paper. In this technique, the amino acids migrate as a solvent flows over a sample of the mixture, in a manner analogous to TLC. In paper chromatography, separation is achieved not by absorption and elution, as in TLC, but rather by successive extractions or *partitioning* between the moving solvent and a fixed layer of water molecules that are bound to the cellulose. For quantitative analysis, the amino acid mixture from hydrolysis is separated by ion-exchange chromatography in automated instruments.

Individual amino acids are identified by their position in the chromatograms. Depending on the solvent system, amino acids with acidic or basic groups may migrate very slowly because of salt formation; in most solvents, amino acids with alkyl chains usually migrate more rapidly than those with polar groups in the side chains. To locate the amino acids in a paper chromatogram, the developed sheet is sprayed with a reagent, **ninhydrin**, which gives a characteristic blue-violet color with all amino acids.

EXPERIMENTS

The objectives of these experiments are to observe the properties of several representative proteins, and to compare the amino acid content by paper chromatography of protein hydrolysates and by specific reactions for amino acids. Development of paper chromatograms requires most of a laboratory period. The paper chromatograms should therefore be started at the beginning of the period; other experiments can then be carried out while the chromatograms are being developed.

1. Paper Chromatography of Protein Hydrolysates

For this experiment, pairs of students should run chromatograms of two different hydrolysis mixtures and reference amino acids, and then exchange and compare results. The protein hydrolysates and amino acids to be used with each, in 0.1 *M* solutions, are as follows:

Casein Hydrolysate	*Gelatin Hydrolysate*
alanine	alanine
aspartic acid	glycine
glutamic acid	glutamic acid
proline	proline

In labeled 10 × 75 mm test tubes or small vials, obtain a few drops each of one of the protein hydrolysates and the four amino acids. In another test tube prepare a mixture of the four reference amino acid solutions by combining a drop of each of the four separate solutions. Prepare capillary tubes for applying the solutions as described for use in TLC (Chapter 7).

Obtain a 10 × 16 cm. rectangle of Whatman #1 filter paper—handle this paper by the edge and avoid touching the center of the paper with your fingers. About 1 cm from the bottom of the paper, along the narrow side, mark with a pencil six evenly spaced dots, with the outer two 1 cm from the edges and the others about 1.5 cm apart. Use the second and fifth spots for the hydrolysis mixture and the mixture of reference amino acids, respectively, and the other four spots for the individual amino acids. Record the sequence of samples in your report sheet and then apply very small drops of the six samples. First practice spotting with a scrap of filter paper; the spots should spread no more than 1 mm in diameter. After applying samples to each lane, allow the spots to dry and then make a second application at the same points.

As a developing chamber, use a 500 ml Erlenmeyer flask. The neck of the flask serves to hold the filter paper in a vertical cylinder during development (Fig. 17.1). Pour into the flask a mixture of 11 ml of n-propanol and 5 ml of concentrated ammonium hydroxide solution. Wipe the neck of the flask, and stopper it with a large cork.

Curl the paper into a cylinder with the long sides butted together and the line of spots around the bottom end (on the inner surface). Line up the edges evenly at the bottom end and hold the cylinder together with a small square of masking tape inside the upper end and projecting a little above the paper with the sticky side out. Insert the cylinder into the neck of the flask and push it down until the spotted end is just immersed (all the way around) in the developing solvent. Press the tape against the neck, cork the flask and allow the development to proceed for 2 hours or until about 15 minutes before the end of the laboratory period.

Remove the paper, mark the solvent front with a pencil and allow the solvent to evaporate. Spray the paper uniformly with ninhydrin spray and place the paper in an oven or over a hot plate for a few minutes. Outline the spots with a pencil, indicate their color and mount the paper on your report sheet.

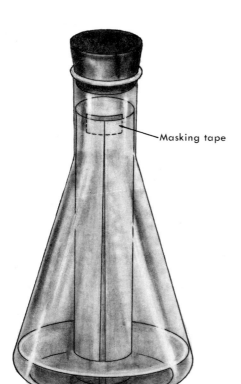

Masking tape

FIGURE 17.1 Paper chromatography of amino acids.

2. Properties of Proteins

a. Solubility. In test tubes place 100 mg samples of casein (use material from Chapter 15, if available), gelatin and egg white (use about 0.5 ml of fresh egg white or 100 mg of dried egg albumin). Add 2 ml of water to each tube, stir thoroughly and record the appearance. Add 1 drop of 2 N NaOH solution, stir and record any changes. Then add 3 to 4 drops of 2 N HCl and record changes.

b. Reaction with Concentrated HNO$_3$. Transfer about half of each of the solutions or suspensions from the solubility tests in Part 2a to another test tube, and to each of these tubes add 0.5 ml of concentrated HNO$_3$. Stir the contents, warm briefly in a hot water bath and record the appearance. Then cool the tubes and add 2 N NaOH solution until the solutions are basic, and record any changes.

c. Acid Hydrolysis. To the remaining solutions or suspensions from Part 2a, add 2 ml of concentrated HCl (in the hood). Place the three tubes in a boiling water bath in the hood and record any changes in color. Loosely cork the tube containing the casein solution and heat in a boiling water bath for 30 to 40 minutes.

Cool the casein solution and add a small amount of decolorizing carbon. Shake and observe the color after the carbon settles; and add a little more carbon if needed to remove most of the color. Filter the solution and add 4 N NaOH in 0.5 ml portions until the solution is basic. Compare the biuret reaction of this hydrolyzed casein solution with that of casein in Part 2d.

d. Biuret Reaction. Place samples of casein, gelatin and egg white (about the same amounts used in Part 2a) in clean test tubes. Add 1 ml of water and a drop of 4 N NaOH and transfer half of each solution to another test tube. To each of these test tubes, and also to the hydrolyzed casein solution from Part 2c, add 10 drops of 1% copper sulfate solution. Record the appearance of the four tubes.

e. Sulfur-containing Amino Acids. To the remaining alkaline solutions from Part 2d (those not used in the biuret test), add 1 ml of 4 N NaOH. In a fourth test tube place about 50 mg of hair clippings or powdered keratin and add 1 ml of 4 N NaOH. Heat the solutions for several minutes in a boiling water bath and record changes in appearance. Cool the solutions, add 1 ml of 5% basic lead acetate solution to each, shake and record the appearance.

Date _____

Name _____

Section _____

REPORT: CHAPTER 17

1. Paper Chromatography

Attach chromatogram

Substance
applied   —   —   —   —   —   —

2. Properties of Proteins

a. Solubility.

	Casein	Gelatin	Egg White
Appearance after water addition			
After NaOH addition			
After HCl addition			

b. Reaction with HNO_3.

	Casein	Gelatin	Egg White
Appearance after HNO_3			
After NaOH			

Which of the proteins contains significant amounts of aromatic amino acids?

c. Acid Hydrolysis.

Color changes on heating with concentrated HCl.

	Casein	Gelatin	Egg White

Which of the proteins contains significant amounts of tryptophan?

d. **Biuret Reaction.**

Color with $CuSO_4$.

			Hydrolyzed
Casein	*Gelatin*	*Egg White*	*Casein*

e. **Sulfur-Containing Amino Acids.**

Appearance after heating with NaOH.

Casein	*Gelatin*	*Egg White*	*Hair*

Appearance after adding lead acetate.

Casein	*Gelatin*	*Egg White*	*Hair*

Rank the four proteins according to the amounts of cysteine (1 highest, 4 lowest).

1.

2.

3.

4.

1. Gelatin is obtained by partial hydrolysis of the native protein collagen, in which a number of amino acids are absent. Two amino acids are present in relatively large amounts — 27% and 15%. From the paper chromatography results with the gelatin hydrolysate, which amino acids are these major components?

2. The reaction of proteins with nitric acid is due mainly to the nitration of tyrosine residues. Write the structure of the product formed by nitration of a tyrosine unit with excess nitric acid, and account for the change in color on making the solution basic.

3. What is the compound responsible for the black color in the lead acetate test for sulfur-containing amino acids? Show the mechanism for the reaction of a cysteine unit with base. (Hint: base hydrolysis of proteins causes extensive *racemization* of the α-amino acids; racemization and the destruction of cysteine result from the same initial step with base.)

eighteen

lipids

The term **lipid** includes all substances in living tissues that are soluble in ether, methylene chloride or similar organic solvents. The most abundant lipids are triesters of glycerol with long-chain "fatty" acids. The triesters or **triglycerides** from animal sources are usually low-melting solids, and are called fats, while those from plants are viscous oils that solidify below 0°. The general structures are the same, and the difference in properties arises from the fact that the vegetable oils contain a larger proportion of unsaturated acid groups, with one, two or three double bonds in the chain. Triglycerides such as beef tallow and other depot fats of animals contain mostly acid chains with one or no double bonds. Fatty acids invariably have chains with an *even* number of carbon atoms, most commonly 16 or 18. A few of the more important fatty acids are listed below; these and several others can occur in any combination in a given triglyceride.

Typical Fatty Acids		*Triglyceride Structure*
palmitic	$CH_3(CH_2)_{14}CO_2H$	$R_1-CO-O-CH_2$
stearic	$CH_3(CH_2)_{16}CO_2H$	$R_2-CO-O-CH$
oleic	$CH_3(CH_2)_7CH=CH(CH_2)_7CO_2H$	$R_3-CO-O-CH_2$
linoleic	$CH_3(CH_2)_4(CH=CHCH_2)_2(CH_2)_6CO_2H$	
linolenic	$CH_3CH_2(CH=CH-CH_2)_3(CH_2)_6CO_2H$	

Butter fat is atypical, since the mixture of acids in the triester contains a significant number of 8-, 10- and 12-carbon chains. The composition of a few fats and oils is given in Table 18.1. These values vary over a fairly wide range, depending on the source of the oil, and there is no one "correct" figure.

TABLE 18.1 TYPICAL ACID COMPOSITION IN FATS AND OILS

Source	Less than 16 Carbons	Palmitic	Approximate Percentages Stearic	Oleic	Linoleic	Linolenic
beef fat	4	30	24	40	2	–
butter	18	30	9	26	4	–
corn oil	1	10	3	34	50	–
cottonseed oil	1	23	1	23	48	–
olive oil		7	3	85	5	–
safflower oil		7		19	70	3
soybean oil		10	2	29	50	6

The degree of unsaturation is important in nutrition, since there is evidence that a high proportion of saturated fats in the diet leads to deposition of cholesterol in blood vessels (atherosclerosis). The unsaturation in a triglyceride is expressed as the "iodine number," which is based on titration of a weighed sample of the fat with iodine monochloride, ICl. The addition of iodine to a double bond is reversible, with a very unfavorable equilibrium constant, but addition of ICl is nearly complete. The "**iodine number**" is calculated as the grams of iodine that would be consumed by 100 g of fat, although the reaction is actually the addition of ICl.

$$I_2 + Cl_2 \longrightarrow 2\,ICl$$

$$R-CH=CHR \xrightarrow{\;ICl\;} R-\underset{\underset{I}{|}}{C}H-\underset{\underset{Cl}{|}}{C}HR$$

A qualitative comparison of the degree of unsaturation in different oils can be made by using a solution of bromine in CCl_4. Since addition to a double bond in a long chain is not as rapid as with a small alkene molecule, excess halogen is used to speed up completion of the reaction. In iodine value determinations, the excess halogen is back-titrated with thiosulfate. In a qualitative comparison with bromine, the amount of excess bromine can be estimated by the intensity of the remaining color.

For use in baking, a solid fat with a broad melting range is desirable, and for this purpose, vegetable oils are "hardened" by hydrogenation. The main purpose of this treatment is to convert the doubly unsaturated linoleic acid units to monounsaturated chains together with some saturated stearic acid units.

$$CH_3(CH_2)_4CH=CH-CH_2-CH=CH(CH_2)_7\overset{O}{\overset{||}{C}}-OR \xrightarrow[Ni]{H_2}$$

$$CH_3(CH_2)_4CH=CH(CH_2)_{10}\overset{O}{\overset{||}{C}}O-R + CH_3(CH_2)_7CH=CH(CH_2)_7\overset{O}{\overset{||}{C}}O-R$$

SAPONIFICATION OF FATS

The term **saponification** means conversion to soap. Alkaline hydrolysis of a fat gives glycerol and the salt of the fatty acid, or soap. The combination of a long alkyl chain and ionic end group imparts surface-active properties, i.e., emulsification of oily droplets in water and penetration of water on a greasy surface. The sodium salts that make up common soap are soluble in water, but the solubility is markedly decreased by a high concentration of ions. Addition of salt precipitates the soap, and this simple technique is used to isolate the soap after hydrolysis.

Each ester group in a fat consumes one mole of base in the saponification. For a given weight of triglyceride, the shorter the alkyl chains, the more ester groups per gram, and the more base required per gram of fat. The number of milligrams of KOH required to saponify 1 g of fat is called the **saponification number.** This value, like the iodine number, is determined in the routine analysis of fats and oils.

CHOLESTEROL

Cholesterol is a lipid present in all plant and animal tissues. It occurs in the body as the free alcohol and also as fatty acid esters; the concentration of "free and combined" cholesterol in blood is about 0.2 mg/100 ml. Cholesterol is the parent substance of a large group of compounds called steroids, which includes the hormones of the adrenal glands and gonads. The common structural unit in steroids is a tetracyclic carbon skeleton with a side chain and various substituents. Steroid hormones such as progesterone are derived in the body from cholesterol by oxidative removal of part of the eight-carbon side chain and minor changes in the functional groups.

Cholesterol

Progesterone

Cholesterol — conformation

CHOLESTERYL ESTERS AND
LIQUID CRYSTAL THERMAL MAPPING

A property of many esters of cholesterol is the formation of **liquid crystals**. This seemingly contradictory term describes a physical state which is liquid in properties and general appearance, but in which molecules are partially ordered or "stacked" in a layered structure. The general requirement for the formation of liquid crystals is a long, flat, relatively rigid molecule. If the molecules are asymmetric, the stack of layers, each containing molecules lined up lengthwise, is slightly *twisted* (Fig. 18.1). Cholesterol esters meet the requirements for a twisted liquid crystalline structure. The carbon skeleton contains five asymmetric centers locked into a rigid skeleton, and the long axis of the ring system can be extended by formation of esters.

Because of the layered arrangement of cholesteric liquid crystals, light striking a film of the substance is **scattered**, analogous to the effect seen in the iridescence of an oil film on water. The color of the scattered light depends on the distance between the layers in the liquid crystal. As the spacing increases, light of longer wavelength, i.e., toward the red end of the spectrum, is scattered. If the spacing is decreased, scattering occurs at shorter wavelengths, toward the blue end of the spectrum. These changes in distance occur with very slight raising or lowering of the temperature.

As a result of the very sensitive temperature dependence, a thin film of a cholesteryl ester, or mixture of esters, in the liquid crystalline state can show a play of color from red to blue, owing to the wavelength of the scattered light, over a temperature range of a few degrees. This property

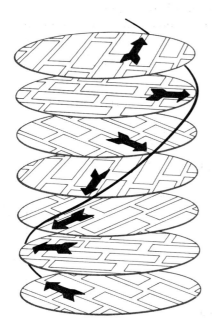

FIGURE 18.1 Schematic diagram of cholesteric liquid crystal.

has been used as a method for "temperature mapping" of a surface. An important application of this technique is in clinical work. Skin temperatures can be very accurately measured, and liquid crystal thermography is now used for routine determination of body temperatures of infants and also in the detection of temperature gradients which may be related to subcutaneous carcinoma.

Mixtures of cholesteryl nonanoate and cholesteryl oleyl carbonate show shifts from red to blue over very narrow temperature ranges. Both of these esters have a liquid crystal phase between the random liquid and the crystalline solid. The pure nonanoate exists as liquid crystals from 77 to 90° and the oleyl carbonate from 20 to 40°. By varying the proportions of the two esters, mixtures can be obtained which show sharp color changes over a few degrees in the region of body temperature. The larger the amount of the nonanoate, the higher the range. Addition of a small amount of a third ester, the benzoate, broadens the temperature range over which the changes occur.

$$R = CH_3(CH_2)_7C-$$

Cholesteryl nonanoate

$$R = CH_3(CH_2)_{17}O-C-$$

Cholesteryl oleyl carbonate

At the higher end of the color-temperature range of these mixtures, the color is blue-violet. As the temperature is lowered the color changes through green and orange to red, and then disappears. Above or below the temperature range of the particular mixture, the scattered light is not visible. The color is caused only by light that is scattered—not transmitted—and the effect is therefore best seen against a black background. For clinical use, a black water-soluble base is applied to the area to be measured, and a dilute solution of the liquid crystal mixture in pentane is then sprayed or brushed on. A kit containing the base coating and liquid crystal mixtures in several temperature ranges is available (Lix Kit). Liquid crystal mixtures are also encapsulated and applied to plastic film for various non-clinical uses.

EXPERIMENTS

1. Unsaturation of Fats and Oils

The objective of this experiment is to compare the degree of unsaturation of several fats and oils. The method is a qualitative adaptation of the iodine number determination.

In labeled 18×150 mm test tubes, place 10 drops of the following oils or melted fats:

Tube a: butter
Tube b: Crisco, Spry or other shortening
Tube c: corn oil (e.g., Mazola)
Tube d: cottonseed oil
Tube e: olive oil
Tube f: safflower oil (or soybean oil)

To each tube add 5 ml of 1 % Br_2—CCl_4 solution. After 10 minutes, compare the color of the test tubes against a piece of white paper and rank them in color intensity. If some of the tubes are nearly colorless and indistinguishable, add another 1 ml of bromine solution to all of the tubes and then rank them again after 5 minutes.

2. Saponification

The objective of this experiment is to prepare a small sample of a soap and observe its properties.

The equipment set-up depends on the method of heating which is used. If a steam bath is available, carry out the reaction in a 150 ml beaker; if a burner must be used, the reaction should be run in a round-bottom flask with a reflux condenser.

In the beaker or flask place 5 g of hydrogenated shortening or vegetable oil and add a solution of 3 g of NaOH in 10 ml of distilled water. Add 10 ml of ethanol, mix and heat the mixture to boiling for about 30 minutes, stirring or shaking occasionally. During the heating period, prepare a salt solution from 30 g of NaCl and 100 ml of water in a 400 ml beaker.

If necessary, add a little water to the saponification mixture to make it fluid, and then pour it into the salt water. Stir the mixture for several minutes and then collect the precipitate on a Büchner funnel with suction. Wash the soap with 5 ml of cold distilled water and then press firmly.

Dissolve about half of the soap in 10 ml of distilled water. In three test tubes place, respectively, 2 ml of the soap solution, 2 ml of water, and 2 ml of water containing a few drops of liquid detergent. To each tube add two drops of mineral oil (or the original vegetable oil). Cork the tubes, shake them vigorously and record the appearance after they have stood for several minutes. To another sample of the soap solution add a few drops of $CaCl_2$ solution and observe any changes.

3. Liquid Crystals

These experiments demonstrate the use of cholesteryl ester liquid crystals for temperature measurement. Solutions of three ester mixtures in pentane solution will be available:

Ester	Mixture 1	Mixture 2	Mixture 3
Nonanoate	55% *green blue*	50% *blue*	40% *purple-black*
Oleyl carbonate	*40°* 35% *blue*	40%	55%
Benzoate	10%	10%	5%

If your lab has black stone bench tops, place a few drops of each ester mixture on the bench. If dark stone tops are not available, put the spots on microscope slides over a black surface. When the pentane has evaporated, smear the spots slightly with your finger. If no colors are visible in one or more of the spots, warm a beaker slightly and set it over the spot. Observe the color changes and the approximate temperature range necessary for each mixture. The temperature-color response of mixtures 1 and 2 can be roughly measured by placing 40° water in the beaker, stirring with a thermometer as the water cools, and observing the color changes through the bottom of the beaker.

The use of liquid crystals in mapping skin temperature can best be illustrated with commercial Lix Kits, but the effects can be seen with the ester mixtures described above. A black base is prepared by dissolving 2 g of polyvinyl alcohol in 25 ml of water and adding 6 g of pharmaceutical-grade carbon black, 5 drops of glycerine and 2 drops of liquid detergent solution.

Swab the back of one hand with rubbing alcohol, shake the black base mixture and apply a thin film with a wooden applicator stick. Allow the coating to dry to a dull black color. Put a few drops of ester mixture 1 on the black area and allow the solution to spread and dry. You may need to experiment with warming or cooling your hand to see color changes. If you can obtain colors only by warming your hand, wipe off the ester with a tissue moistened with pentane and try mixture 2. When the base and solution are properly applied, you should be able to see major veins in the hand traced out in color. Record the appearance of colors under various conditions — for example, after dipping your fingers in ice water. To clean up, wipe off the ester mixture with rubbing alcohol and wash with soap and water.

dyes and dyeing

Dyestuffs include a diverse group of compounds which have in common the property of producing a permanent color on cloth, leather or paper. All of the useful dyes are aromatic compounds with highly delocalized electron systems that absorb light at certain wavelengths. The most important structural types are azo compounds, triarylmethyl cations, and anthraquinones.

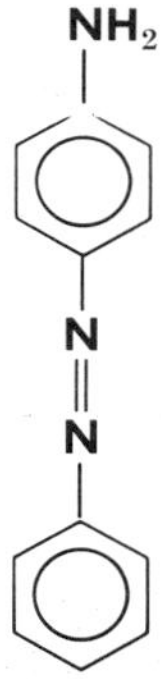

Aniline yellow
(Azo compound)

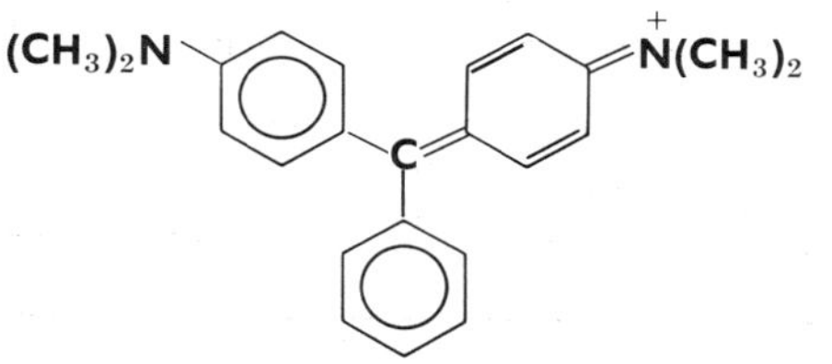

Malachite green
(Triarylmethyl cation)

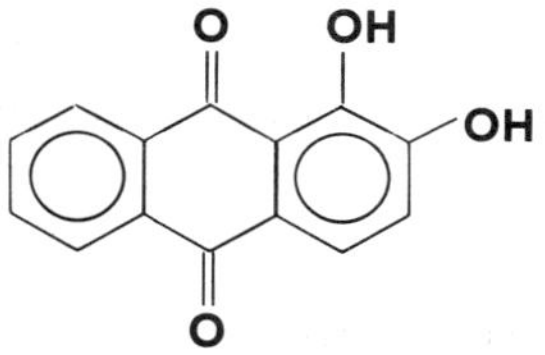

Alizarin
(Anthraquinone)

The chemistry of dyeing involves much more than highly colored compounds. The dyeing process is an interaction between a dye and a fiber, and what counts is the final combination. A beautifully colored substance is of no use as a dye if it does not impart its color irreversibly to the fabric. The most important requirements from the standpoint of the user are pleasing shades and fastness or permanence to washing, air oxidation, perspiration and exposure to strong light.

Another set of considerations comes into the manufacturing process. The dyer must be able to obtain precisely reproducible shade and depth of color. For economic reasons he must also achieve even, level color and at the same time utilize the dye as completely as possible. A major factor in textile dyeing is the large variety of chemical structures present in modern fabrics. These include the natural fibers wool, silk and cotton, and several types of synthetic polymers (Table 19.1).

In dyeing and textile practice, dyes have traditionally been classified

TABLE 19.1 FIBERS AND DYE TYPES

Name	Type	Structural Unit	Type of Dye
wool or silk	protein	$\left(\!\!\begin{array}{c} R\ \ \ O \\ NH{-}CH{-}C \end{array}\!\!{-}OH\right)_n$ R contains CO_2H, NH_2 and OH groups	cationic (Ar_3C^+) or anionic ($-SO_3-$ groups)
cotton or viscose	cellulose	 HO, CH_2, HO, OH, O, $\big)_n$	substantive (azo) $(-N{=}N{-}C_6H_4{-}C_6H_4{-}N{=}N-)$; also vat and mordant
acetate	cellulose acetate	 HO(OAc), CH_2, AcO, OAc, O, $\big)_n$	disperse (anthraquinones and others)
Orlon	acrylic	$\left(\!\!\begin{array}{ccc} CN & CN & CN \\ CH_2{-}CH{-}CH_2{-}CH{-}R{-}CH_2{-}CH \\ & SO_3^- \end{array}\!\!\right)_n$	cationic
Dacron	polyester	$\left(\!\!\begin{array}{c} O\ \ \ \ \ \ \ O \\ C{-}\bigcirc{-}C{-}O{-}CH_2CH_2{-}O \end{array}\!\!\right)_n$	disperse
nylon	polyamide	$\left(\!\!\begin{array}{c} O\ \ \ \ \ \ \ \ \ O \\ NH(CH_2)_6NH{-}C{-}(CH_2)_6{-}C \end{array}\!\!\right)_n$	anionic, also disperse

according to the *dyeing process* or the *fiber* that is dyed. For example, the terms "acid" and "basic" refer to dyes that are applied from an acid or from an alkaline dye bath. Direct dyes are those that can be applied to cotton. Disperse dyes are applied from a *suspension* rather than from a solution of the dye. Vat dyes are applied by using a soluble, reduced form of the dye in the bath and then oxidizing it to an insoluble, colored form on the fabric. All of these classes of dyes can include various structural types. Thus, a disperse dye or an acid dye may have any type of chromophoric (color-producing) structure, provided that the dyeing process involves a suspension of the dye or an acid solution, respectively.

Acid and basic dyes are fixed to the fiber by ionic attraction, and the terms anionic and cationic are more descriptive and meaningful. Typical "acid" dyes contain $-SO_3^-$ Na^+ groups, and combine in acid solution with a fabric containing $-NH_2$ groups. The dyeing process is actually an ion exchange in which the $-NH_3^+$ cation of the fiber replaces Na^+ in the dye. In "basic" dyes the dye molecule has a *cationic* structure and combines with anionic centers in the fiber.

Silk and wool contain both cationic NH_3^+ groups and anionic CO_2^- groups in the amino acid side chains, and are dyed with either type of ionic dye. Synthetic polyacrylic fibers contain SO_3^- groups incorporated

Acid Dyeing:

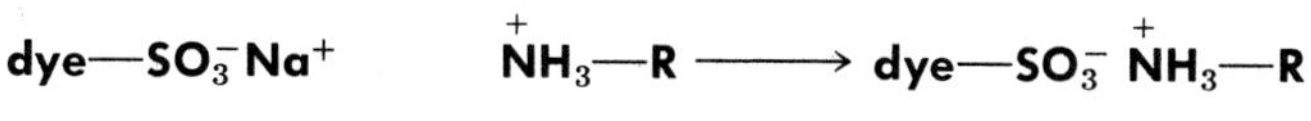

$$\text{dye—SO}_3^-\text{Na}^+ \qquad \overset{+}{\text{NH}_3}\text{—R} \longrightarrow \text{dye—SO}_3^- \ \overset{+}{\text{NH}_3}\text{—R}$$

      Anionic dye   Cationic   Dyed fiber

                                      fiber in
                                      acid solution

Basic Dyeing:

    Cationic dye   Anionic fiber in   Dyed fiber

                               neutral or
                               basic solution

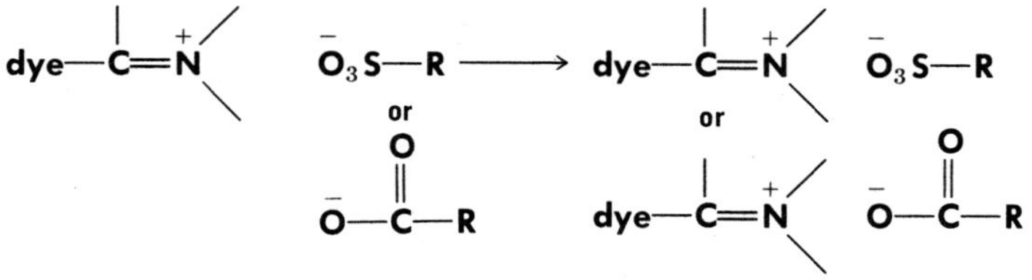

in the polymer to provide dye sites. Nylon can be prepared with either —NH_3^+ end groups or —CO_2^- end groups in excess; most nylon fabrics contain excess —NH_3^+ groups and can therefore be dyed with anionic ("acid") dyes.

Cotton and modified cellulose (viscose, rayon) do not give good fastness (permanence) with simple anionic or cationic dyes. An important group of dyes that are direct or substantive for cotton are compounds in which azo groups occur at a distance that corresponds to the length of the repeating hexose unit in the cellulose chain. The dye is bound to the fiber by a series of hydrogen-bonding attractions. A single azo group is insufficient for dye fastness; "dis-azo" compounds, with two groups linked to a diphenyl system, provide excellent substantive dyes.

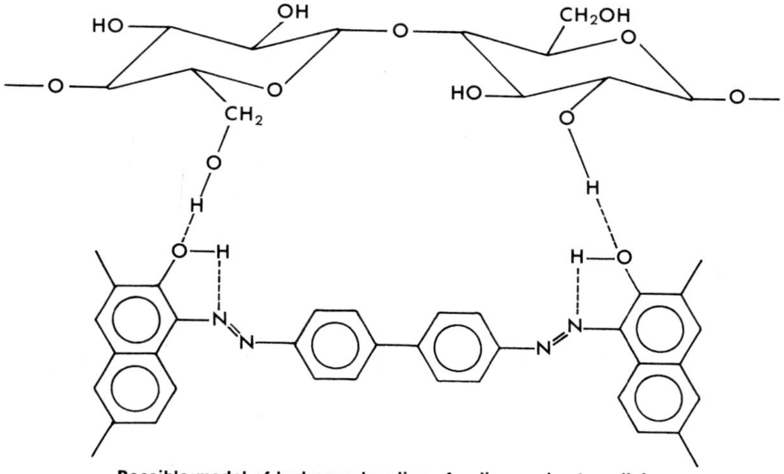

Possible model of hydrogen-bonding of a dis-azo dye to cellulose

Polyester and cellulose acetate fibers have the least affinity for adsorption of dyes from solution, and these fibers are dyed by the disperse process. The fabric is heated in a suspension of an insoluble dye and a "carrier" which enhances penetration. The dye becomes dispersed in the hydrophobic fiber. Since there is no bonding to the fiber, disperse dyes are often less permanent than other types.

An important development in textile practice is the use of mixed and blended fibers. In dyeing these materials, the differential binding of dyes on the individual fibers provides textured colors. The blended fabric is treated with a mixture of dyes, and the *lack* of affinity of a fiber for certain dyes is just as important as its affinity for others. A blended fiber, or a fabric woven with different fibers, can be cross-dyed with a mixture of dye types. In this way, patterned colors can be achieved in a single dyeing operation.

An enormous range of dyes is available to cover the diverse requirements for colors and fabrics. The dyeing process is determined by the type of fabric and the dye, but a number of factors contribute to the final result. The concentration of dye, the presence of electrolytes and the temperature are important variables in the dye bath. The degree of wetting and swelling of the fiber, the duration of the dyeing and the addition of surface-active compounds are other parameters that affect the overall outcome. Because of the large number of factors that can be varied and the complicated interactions between dye and fiber, textile dyeing is a highly empirical practice, requiring skill and long experience.

CHEMISTRY OF AZO DYES

The preparation of azo dyes involves two reactions — diazotization and coupling. Both reactions are very simple operations which are carried out in aqueous solution. In diazotization, an aromatic amine is converted to a diazonium ion with nitrous acid.

$$NaNO_2 + HCl \longrightarrow HO{-}NO + NaCl$$

Sodium nitrate

$$HO{-}NO + H^+ \longrightarrow H_2\overset{+}{O}{-}NO \rightleftarrows H_2O + \overset{+}{N}{=\!=}O$$

$$ArNH_2 + \overset{+}{N}{=\!=}O \longrightarrow Ar{-}N{=\!=}N^+ + H_2O$$

Diazonium ion

The coupling reaction is an electrophilic substitution of a phenol, naphthol or aromatic amine to give an azo compound. The electrophile is the ArN$=$N$^+$ ion, and substitution occurs at positions *ortho* or *para* to the —OH or NH$_2$ group. To obtain soluble azo dyes, a naphthol containing

—$SO_3^- Na^+$ groups is used for coupling. A typical reaction is the coupling with the salt of 2-naphthol-3,6-disulfonic acid ("R-salt"). The resulting azo compound can be used as an anionic or "acid" dye.

"R-salt" Anionic azo dye

If a naphthol without solubilizing —SO_3Na groups is employed, the azo compound is insoluble, and can be used as a disperse dye.

In certain dyeing applications, particularly textile printing, the azo coupling reaction is carried out *on the fabric*. The cloth is first immersed in an alkaline solution of the naphthol, or a paste of the naphthol is applied in a pattern. The fabric is then immersed in the diazonium solution, and the dye is formed on the surface of the fiber.

EXPERIMENTS

The experiments in this chapter illustrate several aspects of dye chemistry and the dyeability of various fibers. In a brief laboratory experiment it is not possible to obtain optimum dyeing conditions or to evaluate the permanence of a dye. However, the relative affinities of fibers for dye types can be observed by simply exposing samples of various fibers to different dyes under the same conditions. Multifiber cloth containing bands of several different fibers provides a simple means for evaluating and comparing dye affinities. One standard multifiber cloth, number 10, has six bands in the following sequence: wool, Orlon, Dacron, nylon, cotton and acetate.

The amines to be used in these experiments are listed in Table 19.2. In Parts 1 and 2, you will select one monoamine and one diamine from the list. The amines are available as 0.05 N solutions in 0.1 N HCl—each solution contains 0.05 mmole of monoamine or 0.025 mmole of diamine and 0.1 meq of HCl per ml. These solutions are more dilute than those normally used for diazotization, but the rather high dilution is advantageous for an experiment of this type, and permits the reactions to be carried out rapidly by a standard procedure.

TABLE 19.2

	Name	Mol. wt.	mg/ml
Monoamines			
	o-anisidine	123	6.2
	m-nitroaniline*	138	6.9
	p-nitroaniline*	138	6.9
Diamines			
	benzidine	184	4.6
	dianisidine	244	6.1
	o-tolidine	212	5.3

*The solutions of these amines contain 0.2 N HCl.

GENERAL PROCEDURE FOR DIAZOTIZATION

Place 10 ml (0.5 meq) of a solution of mono- or diamine in a test tube and chill several minutes in an ice bath. Add 1 ml of 0.5 N NaNO$_2$ solution to the amine and note any change in appearance. The diazonium solutions must be kept at 0° in an ice bath until they are used.

1. Dyeing on the Fiber

Obtain two 2 × 3″ swatches of cotton and moisten them partially with a 1.0 M solution of β-naphthol (the solution contains 0.144 g [1 mmole] of naphthol and 2 meq of NaOH per ml). Let several drops of solution splash in a pattern, or twist the cloth into a loop and partially immerse this in the solution. Blot the swatches with a paper towel and allow them to dry.

Prepare diazonium solutions from a monoamine and a diamine as described in the general procedure. Pour the solutions into small beakers and immerse one of the dried naphthol-treated patches in each solution. Handle the wet patches with a glass rod—not your fingers. (**CAUTION:** Avoid handling the dyed cloth in any of these experiments. The dyes are not harmful, but skin or fingernails will remain stained for several days.) Rinse the patches thoroughly in running water, blot them with a paper towel and, when dry, attach them to your report.

2. Comparison of Mono- and Dis-azo Dyes in Neutral and Acid Solution

Select a monoamine and a diamine different from those used in Part 1, obtain 10 ml of each and diazotize according to the general procedure. If the monamine used is one of the nitroanilines, add 1 ml of 1 N NaOH to the diazonium solution to neutralize the excess acid. To each diazonium solution add 1 ml of a 0.5 M solution of "R salt" (this solution contains 0.174 [0.5 mmole] of R salt and 1 meq of NaOH per ml).

Place the dye solutions in a water bath at 70 to 80°, and in each solution immerse a 1″ strip of number 10 multifiber cloth and a 1 × 2″ patch of cotton. Keep the solutions in the bath about 5 minutes and then remove the tubes. Transfer the cloth samples with a glass rod to clean beakers and rinse the samples thoroughly with water and blot on a towel.

To the dye solutions add 2 ml of 2 N HCl (or 1 ml of 4 N acid) and repeat the dyeing in each bath with a cotton patch and a strip of multifiber cloth. After 5 minutes, remove, rinse and blot the samples. Save the dis-azo dye bath (from the diamine) and discard the other dye solution.

Compare the intensity of color of the cotton strips dyed with the mono- and the dis-azo dyes, and also compare the intensity of the color in the various multifiber bands for the two dyes in neutral and in acid solutions.

3. Disperse and Cross-Dyeing

Prepare a diazonium solution from 10 ml of m-nitroaniline. Neutralize the diazonium solution with 1 ml of 1 N NaOH and then add, dropwise with stirring, 1 ml of β-naphthol solution. Allow the solution to warm to room temperature and stir and rub, if necessary, to obtain a finely divided suspension of solid. Add about 5 mg of biphenyl and a few drops of surfactant solution to the dye mixture and immerse a strip of multifiber cloth.

(Alternatively, if available, 10 ml of commercial yellow disperse dye suspension can be used instead of the dye prepared from m-nitroaniline.)

Heat the dye bath in a 90 to 100° water bath for 10 to 15 minutes, and then cool and remove, rinse and blot the sample. Record the relative intensities of the bands in the multifiber sample.

Combine the remaining dye suspension and the remaining acid solution of dis-azo dye from Part 2, and repeat the dyeing with another multifiber strip at 90°. If available, a sample of checked multifiber cloth (number 7403) can be dyed also. Remove, rinse, blot and record the appearance of the dyed sample.

4. Cationic Dyes

Obtain 10 ml of a 0.1% solution of malachite green and immerse a strip of multifiber cloth. Hold the solution in an 80 to 90° bath for several minutes and then rinse, blot and record the relative intensities of color in the bands.

REPORT: CHAPTER 19

1. Dyeing on Fiber

Monoamine used:

Diamine used:

Structures of dyes formed:

_____   _____

Attach samples of dyeings:

2. Comparison of Mono- and Dis-azo Dyes in Neutral and Acid Solution

Monoamine used:

Diamine used:

Structures of dyes:

Mono-azo   **Dis-azo**

a. Cotton dyeing with mono- and dis-azo dyes.
   Attach samples:

Mono-azo — neutral bath	Dis-azo — neutral bath	Mon-azo — acid bath	Dis-azo — acid bath

Which dye has the greater affinity for cotton?

What differences were observed between neutral and acid dye baths with the cotton samples?

b. Affinities for various fibers.
   Attach samples of multifiber strips (place wool band on top).

Mono-azo— neutral bath	Dis-azo— neutral bath	Mono-azo— acid bath	Dis-azo— acid bath

Which fibers are dyed effectively in the four samples?

Mono-azo—neutral:

Dis-azo—neutral:

Mono-azo—acid:

Dis-azo—acid:

For which fibers does the pH of the bath have the greatest effect?

3. Disperse and Cross-Dyes

Attach samples of dyeings:

Disperse	Cross dyeing

Which fibers were effectively dyed in the disperse process and which were not?

4. Cationic Dyes

Attach sample of dyeings:

Which fibers were effectively dyed by cationic dyes and which were not?

QUESTIONS

1. What type of dye process would be best suited for polypropylene fibers $\left(\text{polypropylene is } \left[\text{CH}_2\text{—}\overset{\overset{\displaystyle \text{CH}_3}{|}}{\text{CH}}\right]_n\right)$?

2. What generalizations can you make concerning color and constitution; i.e., dye structure, from your observations in this experiment?

infrared and nmr spectra

Light and other electromagnetic radiation such as infrared (heat) rays are absorbed by organic molecules when the frequency of the radiation corresponds to some change in energy level in the molecule. The extent of absorption and the frequency of the radiation are measured in a spectrometer, and a spectrum, such as that shown in Figure 20.1, is recorded.

INFRARED SPECTRA

The position of an absorption band in an infrared spectrum can be expressed in terms of either the **frequency** or the **wavelength** of the radiation. Frequency and wavelength have a reciprocal relationship; i.e., as one increases, the other decreases. The wavelength, λ, of infrared radiation is in the region 2 to 15 **microns** (a micron, μ, is 10^{-4} cm). Frequency is expressed as the **wave number**, ν, which is the *reciprocal* of the wavelength in cm, or **cm^{-1}**. Thus:

$$\nu \text{ (cm}^{-1}) = \frac{1}{\lambda \text{ (cm)}} = \frac{1}{\lambda \text{ (}\mu) \times 10^{-4}} = \frac{10000}{\lambda \text{ (}\mu)}$$

$$\text{and } \lambda \text{ (}\mu) = \frac{10000}{\nu \text{ (cm}^{-1})}$$

Examples. (a) $\lambda = 6.0$ (b) $\nu = 2000$ cm^{-1}

$$\nu = \frac{10000}{6} = 1670 \text{ cm}^{-1} \qquad \lambda = \frac{10000}{2000} = 5.0 \ \mu$$

Infrared spectrophotometers can be designed with the chart calibrated in either wave number in cm^{-1} or wavelength in μ, and both types are used. Values for infrared vibrations are therefore given both in cm^{-1} and in μ.

The energy levels responsible for infrared absorption (wave numbers in the range 4000 to 650 cm^{-1} or wavelengths from 2.5 to 15.4 μ) are those involved in atomic vibrations in covalent bonds. The most important vibrational modes are **stretching** of bonds and **bending** of atomic groupings. The frequencies depend on the mass of the atoms, e.g., —C—H, and the

199

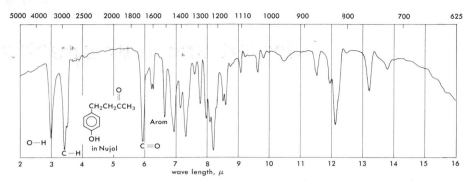

FIGURE 20.1 Infrared spectrum of 1(4-hydroxyphenyl)-3-butanone.

type of bond—single, double or triple. Table 20.1 shows the most common types of bonds involved in several regions of the spectrum.

In this discussion, only the bands of importance in connection with the unknown experiment (Chapter 21) will be discussed. The most useful bands for functional group identification are those due to O—H or N—H and C=O stretching. These bands, together with the C—H stretching around 3000 cm^{-1} (3.33 μ) are the most prominent ones in the 4000 to 1600 cm^{-1} (2.5 to 6.25 μ) region of the spectrum. With a few exceptions, most of the absorption bands in the region below (to the right of) 1600 cm^{-1} (6.25 μ) cannot be assigned to a specific bond or atomic grouping, but the numerous bands in this region provide a "fingerprint" which uniquely characterizes the compound.

C—H and C—C Bonds

In aliphatic compounds, C—H stretching vibration usually gives a strong band at 3000 to 2800 cm^{-1} (3.3 to 3.6 μ), which often appears as two partially resolved peaks. (C—H bending leads to a band around 1400 cm^{-1} or 7.15 μ.) Aromatic C—H stretching absorption occurs at slightly higher frequency, 3100 to 3000 cm^{-1} (3.2 to 3.3 μ), and in compounds with both aliphatic and aromatic C—H, the two stretching vibrations can often be seen in the spectrum. Aromatic ring vibrations cause strong bands around 1600 to 1500 cm^{-1} (6.25 to 6.65 μ); most spectra contain bands due to C—C bonds in the 1300 to 1100 cm^{-1} (7.7 to 7.1 μ) region.

TABLE 20.1

ν, cm^{-1} (λ, μ)	Type of Bond
3600–3200 (2.78–3.12)	O—H or N—H stretching
3200–2900 (3.21–3.45)	C—H stretching
2400–2000 (4.16–5.00)	C≡N or C≡C stretching
1800–1660 (5.55–6.00)	C=O stretching
1600–650 (6.25–15.4)	C—C, C—O and C—N stretching and many types of bending

Infrared spectra of solid compounds are often measured by using a suspension or mull of the compound in the straight-chain alkane Nujol. In these **Nujol mull** spectra, strong peaks at 2940 to 2860 cm^{-1} (3.4 to 3.5 μ) and 1450 cm^{-1} (6.9 μ) and a weaker peak at 1370 cm^{-1} (7.3 μ) due to the Nujol are always present.

0—H Bonds

The characteristic infrared band due to O—H stretching in **alcohols** or **phenols** appears around 3600 cm^{-1} (2.78 μ) in dilute solutions, but in the spectra of pure liquids or solids, which is the case with the spectra that will be used in Chapter 21, the band is broad and is shifted to 3400 to 3200 cm^{-1} (2.95 to 3.12 μ) because of hydrogen bonding. In **acids**, the OH group of —CO$_2$H is even more broadened.

N—H Bonds

The bands due to N—H stretching in **amines** also occur around 3500 to 3300 cm^{-1} (2.86 to 3.04 μ), overlapping the OH region. The N—H bands are sharper and less intense than OH; primary amines, RNH$_2$, usually have two bands separated by about 70 cm^{-1}.

C=O and C—O Bonds

The carbonyl stretching band is very prominent in the spectra of **aldehydes, ketones, acids** and **esters,** and the position varies slightly with the type and environment. In aliphatic compounds the band appears at 1730 to 1710 cm^{-1} (5.77 to 5.85 μ). In compounds with the C=O group conjugated with a double bond or aromatic ring, the band appears around 1710 to 1690 cm^{-1} (5.85 to 5.92 μ). The carbonyl band is strong and sharp; in compounds with more than one C=O group, separate bands are usually seen.

In **esters,** an additional characteristic band due to the CO—O stretching frequency is present in the range 1250 to 1150 cm^{-1} (8.0 to 8.7 μ). This is also a strong band, but it may be difficult to identify because of numerous other bands in this region.

NMR SPECTRA

Nuclear magnetic resonance spectroscopy depends on differences in energy levels in certain atomic nuclei which possess nuclear spin. The most important case in organic chemistry is the hydrogen nucleus, or proton. In a powerful magnetic field, the protons in a sample behave as

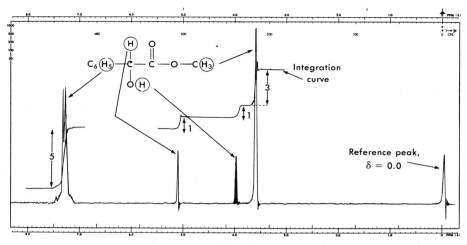

FIGURE 20.2 Nmr spectrum of methyl mandelate.

tiny magnets which line up in one of two directions. The absorption of low-energy radiation (radio frequency region) causes the direction, or spin, to reverse, and this change is detected and recorded. Since protons in different environments within a molecule experience slightly different magnetic fields, they absorb at different frequencies, and the result is an nmr spectrum such as that in Figure 20.2.

The nmr spectrum provides three kinds of information about the protons in a molecule:

1. The position of the signal for each type of proton (same environment) along the horizontal scale depends on the atom to which those protons are bonded and also the neighboring atoms. The position is called the δ value or **chemical shift,** and is expressed as parts per million (ppm) from a reference peak which appears at zero at the right-hand end of the spectrum.

2. The relative size or area of the peaks is given by the **integration curve.** This is the upper line on the spectrum, with a step corresponding to each signal. The heights of the steps are proportional to the number of protons contributing to the peaks, i.e., the number of each type of proton.

3. **Splitting** of the signal from a given proton or group of equivalent protons indicates the presence of protons on *neighboring* atoms.

Chemical Shifts and Peak Areas

The positions of the four proton groupings that are labeled in Figure 20.2 illustrate the distribution of chemical shifts of protons in different environments. *Each proton in the same environment has the same chemical shift.* Thus, the peak at 3.6 ppm in Figure 20.2 due to the CH_3 group arises from all three protons in the group, and the integration step for this peak is three times higher than those of the —C—H or —O—H peaks.

The peak due to the C_6H_5 protons at 7.1 to 7.3 ppm is broadened because the five aromatic protons are not completely equivalent. The integration step for the entire group is five times higher than that of the one-proton peaks.

A list of chemical shifts covering most of the types of protons that will be encountered in the unknowns in Chapter 21 is given in Table 20.2. From the data in this table, it can be seen that peaks for aliphatic protons appear at progressively larger δ values (shifted to the *left*) in $CH_3—>—CH_2—>CH$. An *adjacent* double bond or aromatic ring, as in the case of a benzyl proton, $H—\overset{|}{\underset{|}{C}}—C_6H_5$, causes a further change in the same direction. An adjacent electronegative atom such as oxygen or halogen, as in $H—C—O—$, causes a still larger shift to the left. These "deshielding" effects are caused by inductive withdrawal of electron density from the CH_3, CH_2 or CH, and decrease as the distance between the substituent and the proton in question increases. These factors can be seen in the following examples:

$$\begin{array}{cccc} & \overset{1.61}{\underset{|}{H}} & \overset{H}{|} & \\ 0.88 & | & 1.04\;| & \\ CH_3 & —C— & CH_2—C— & CH_3 \\ & \underset{|}{CH_3} & \underset{|}{CH_3} & \end{array} \qquad \begin{array}{ccc} 0.92 & 1.57 & 3.58 \\ CH_3— & CH_2— & CH_2—OH \end{array}$$

Protons attached *directly* to unsaturated carbon and especially to an aromatic ring are "deshielded" by another type of effect and appear at δ values in the range 5 to 8. In monosubstituted benzene derivatives such as $C_6H_5—CH_3$ or $C_6H_5—Cl$, the *o*, *m* and *p* protons are not exactly equivalent, but they sometimes appear as a single peak which may be somewhat broad (Fig. 20.2). In other cases, such as $C_6H_5—CO—X$, the *o* protons appear at a somewhat higher δ values (7.5 to 8.0 ppm), and the aromatic proton peak is therefore a broad multiplet. Electron-releasing substituents such as —OH or —NH_2 cause the *o* and *p* protons to appear at a smaller value (6 to 7 ppm).

Peaks due to protons in —OH or —NH— groups have a variable position in the spectrum and are sometimes, but not always, very broad. These peaks can be recognized by the fact that they disappear due to H=D exchange when D_2O is added to the sample before the spectrum is recorded. In the spectra of unknowns in Chapter 21, peaks due to —OH or —NH— are *shaded dark* to indicate that they are exchangeable, as shown for the OH peak in Figure 20.2.

Peak Splitting

The splitting of peaks by neighboring protons is caused by interaction or **coupling** of the nuclear spins. In a system $H_A—\overset{|}{\underset{|}{C}}—\overset{|}{\underset{|}{C}}—H_B$, the spin of

TABLE 20.2 CHEMICAL SHIFTS OF PROTON GROUPINGS*

Group	Shift, δ ppm	Group	Shift, δ ppm
Methyl Groups (CH₃)		*Methylene Groups (RCH₂—)*	
CH₃—R	0.8–1.2	R—CH₂—R	1.1–1.5
CH₃—CR=C$\diagdown^{\diagup}$	1.6–1.9	R—CH₂—Ar	2.5–2.9
CH₃—Ar	2.2–2.5	R—CH₂—COR	2.6–2.9
$\overset{\text{O}}{\overset{\|}{\text{CH}_3\text{C}}}$—R	2.1–2.4	R—CH₂OH	3.2–3.5
$\overset{\text{O}}{\overset{\|}{\text{CH}_3}}$—CAr	2.4–2.6	R—CH₂OAr	3.9–4.3
$\overset{\text{O}}{\overset{\|}{\text{CH}_3\text{COR}}}$	1.9–2.2	R—CH₂—$\overset{\text{O}}{\overset{\|}{\text{OCR}}}$	3.7–4.1
$\overset{\text{O}}{\overset{\|}{\text{CH}_3\text{C}}}$—OAr	2.0–2.5		
CH₃—N$\diagdown^{\diagup}$	2.2–2.6		
CH₃—OR	3.2–3.5		
CH₃—OAr	3.7–4.0	*Methine Groups (R₂CH—)*	
$\overset{\text{O}}{\overset{\|}{\text{CH}_3\text{—OCR}}}$	3.6–3.9	R₃CH	1.4–1.6
$\overset{\text{O}}{\overset{\|}{\text{CH}_3\text{OCAr}}}$	3.7–4.0	R₂CHOH	3.5–3.8
		Ar₂CHOH	5.7–5.8
Unsaturated Groups			
RCH=C$\diagdown^{\diagup}$	5.0–5.7	*Other Groups*	
		ROH	3–6
H—Ar	6.0–7.5	ArOH	6–8
(ortho H on aromatic ring with C=O)	7.5–8.0	RCO₂H	10–12
		RNH=	2–4
R—$\overset{\text{O}}{\overset{\|}{\text{C}}}$—H	9.4–10.4		

*R = saturated carbon (CH₃—, —CH₂—, —CH, —C—).
Ar = aromatic ring.

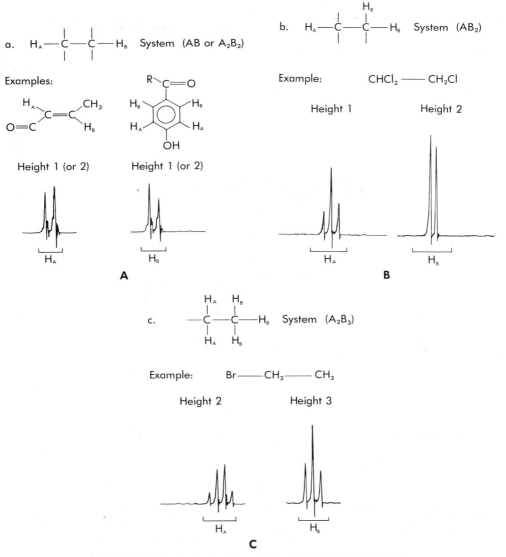

FIGURE 20.3 Typical splitting patterns due to spin coupling.

H_A affects the magnetic field around H_B, and *vice versa*. As a result, the peak due to H_A is split into a **doublet** by H_B, and the peak due to H_B is split into a doublet by H_A. As a general rule, the signal for a proton or group of equivalent protons is split into **n + 1** separate peaks or lines by **n** adjacent protons; the separation of the lines is usually about 0.1 ppm.

Some characteristic patterns resulting from spin coupling are shown in Figure 20.3. One of the most common is that of an ethyl group, CH_3—CH_2. As seen in Figure 20.3, the —CH_2— peak is a **quartet**, split by $(3 + 1)$ CH_3 protons, and the CH_3— peak is a **triplet**, split by $(2 + 1)$ CH_2 protons. It is important to realize that the CH_3 triplet is *one* signal split into three

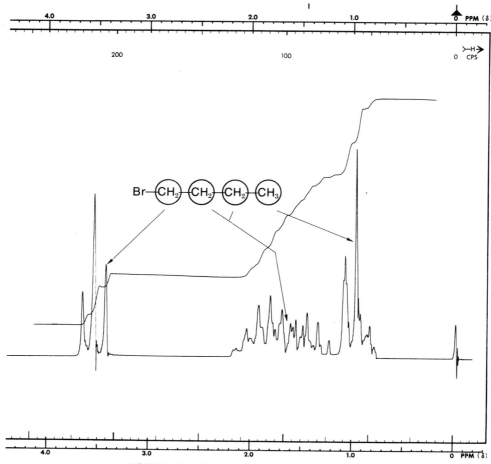

FIGURE 20.4   Nmr spectrum of *n*-butyl bromide.

lines; the three lines are *not* due to the fact that the peak arises from three protons. Another frequent pattern is that shown in Figure 20.3a, due to the protons on a *p*-disubstituted benzene ring. Each of the doublets is due to a pair of protons *o* to one substituent. The lines in multiplet peaks are not the same height; the higher line is always the inner one, closest to the group to which the multiplet is coupled.

It should be kept in mind that the δ values in Table 20.2 will often represent the center of a multiplet peak, and not a single line. In compounds containing several CH_2 groups, such as $CH_3CH_2CH_2CH_2Br$ (Fig. 20.4), the peaks due to the central CH_2 groups form a complex multiplet owing to the extensive spin coupling and closeness of the chemical shifts. The CH_2 group adjacent to Br is well-separated, and is an easily recognizable triplet, due to coupling with the neighboring CH_2. The grouping of peaks at 0.8 to 1.2 (3 protons on the integral curve) is due to the CH_3 protons; this distorted triplet is characteristic of longer straight-chain alkyl groups.

identification of unknowns

An important part of experimental organic chemistry is the characterization and identification of compounds which may be encountered among reaction products or from natural sources such as a plant extract. The objective of this experiment is to learn the approach that is used in identifying a compound, and to carry out the steps in the process using unknowns. The data that are used include spectra, physical properties, solubility classification, chemical tests and, finally, conversion of the unknown to a derivative.

The initial steps will be described first, and these will be performed with some general unknowns to provide experience in the observations. You will then receive samples of compounds from among those listed in the tables at the end of the chapter, and carry out the identification.

INITIAL STEPS IN IDENTIFICATION

Spectra

The two most informative and useful spectroscopic methods for characterizing organic compounds are infrared and nmr spectra. The type of information available from these spectra is discussed in Chapter 20. You will receive copies of the infrared and nmr spectra of your unknowns, and one of the first steps in the identification is the interpretation of these spectra to obtain information on the presence or absence of functional groups, and the type of proton groupings. The initial interpretation may be only partial and, after other data are obtained, further study of the spectra may reveal more complete information. The infrared spectrum, together with the solubility classification of the unknown, will in most cases indicate the type of compound. The nmr spectrum and the melting points of the unknown and of a derivative should permit you to pinpoint your compound in the tables.

Physical Properties

In the days before spectroscopy, the main physical constant by which chemists characterized compounds was the boiling point or the melting point. These properties are less crucial now, but they are still useful and will help to narrow the range of possible identities of the unknown.

Boiling Point

The boiling point of a liquid depends on the molecular weight and the functional groups, and for relatively small molecules with a given functional group, the boiling point gives an indication of the approximate size of the molecule. For compounds which boil above 180 to 200°, so many possibilities exist that the boiling point is less characteristic; moreover, boiling is often accompanied by decomposition, and distillation is carried out at reduced pressure. For these reasons, boiling points above the range of 180 to 200° are not very useful. If the boiling point of your unknown is above 180°, you will be given this information, and the boiling point determination should be omitted.

Procedure. For more volatile liquids, the boiling point can be determined by distilling the unknown in a small flask (Fig. 21.1).

Clamp the flask over an iron ring. On the ring place an asbestos board with a small hole so that excessive heating is avoided. Fit the thermometer in place with the rubber sleeve used in the distillation set-up shown in Figure 4.5; slip a chilled test tube over the side arm as a receiver. Control the rate of heating to permit time for the neck of the flask and thermometer to be heated by the vapor before steady distillation begins.

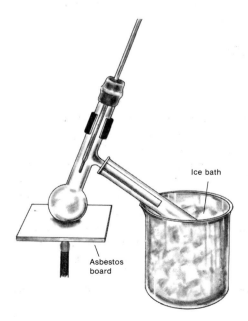

Ice bath

Asbestos
board

FIGURE 21.1 Small scale distillation of unknown.

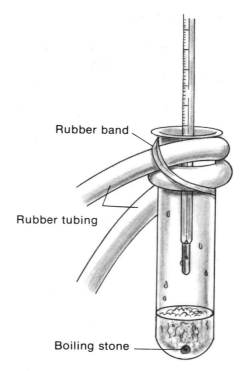

FIGURE 21.2 Reflux set-up for boiling points.

Rubber band

Rubber tubing

Boiling stone

Procedure. Another set-up for determining the boiling point is shown in Figure 21.2. Clamp an 18 × 150 mm test tube resting on a wire gauze or asbestos board and wrap two turns of rubber tubing around the upper end to serve as a condenser. Hold the tubing in place with a rubber band, connect one end to a water outlet and place the other end in the drain. Place 1 to 2 ml of the liquid in the test tube and clamp a thermometer with the bulb below the level of the tubing about 1″ above the liquid. Pass a slow stream of water through the tubing. Heat the liquid with a very small flame until gentle refluxing occurs on the walls and thermometer. When the temperature becomes steady, record this value as the boiling point.

Melting Point

The melting point of a solid does not provide much information about the structure or size of the compound, but it is a characteristic physical constant which can be readily determined and compared with a value previously recorded for the compound. For comparison purposes, the melting point is generally more reliable than the boiling point of a liquid, and it can be observed with a very small sample.

For solid unknowns, the melting point should be determined in the usual way, as described in Chapter 2. In comparing your value with melting points in the tables, a "leeway" of several degrees should be allowed for possible errors in your determination or in the value recorded in the table.

Solubility Classification

The solubility of an organic compound in water or aqueous acid or base can provide evidence for the presence of several important functional groups, as indicated in the following chart.

Medium	Some Solubility or Complete Miscibility
water	alcohols, amines or carboxylic acids with 4 carbons or less
5% NaHCO$_3$	carboxylic acids
5% NaOH	carboxylic acids and phenols
5% HCl	amines

Solubility in water denotes a rather high ratio of polar group to carbon chain, i.e., a low molecular weight compound containing an OH, NH$_2$ or CO$_2$H group, or a larger molecule containing more than one such group. The presence of an acidic CO$_2$H or basic NH$_2$ group in a water-soluble compound can be detected by a low or high pH, respectively, of the solution.

Compounds which are insoluble in water generally dissolve to a significant extent in aqueous acid or base if they form an ionic species. The solubility of carboxylic acids ($K_A = 10^{-3}$ to 10^{-5}) and phenols ($K_A = 10^{-9}$ to 10^{-10}) in aqueous hydroxide is due to the formation of the carboxylate or phenoxide, since they are much *stronger acids* than water ($K_A = 10^{-14}$), and the acid-base equilibria lie far to the right:

$$RCO_2H + OH^- \rightleftharpoons RCO_2^- + H_2O$$

$$ArOH + OH^- \rightleftharpoons ArO^- + H_2O$$

Carboxylic acids, but not phenols, are also stronger than carbonic acid ($K_A = 10^{-7}$), and they are therefore soluble also in NaHCO$_3$ solution:

$$RCO_2H + HCO_3^- \longrightarrow RCO_2^- + H_2O + CO_2$$

The solubility of amines in dilute aqueous acid similarly reflects the fact that they are *stronger bases* than water, and are converted to an ammonium ion:

$$RNH_2 + H_3O^+ \rightleftharpoons RNH_3^+ + H_2O$$

Amines are the only common class of organic compounds which are protonated in dilute aqueous acid.

Procedure. Place two drops of a liquid unknown, or an equivalent amount of a solid, in a test tube and add about 0.5 ml of water. If the compound dissolves completely or partially (a swirling appearance in the water, due to change in the refractive index, is an indication of some solubility), consider it soluble and proceed no further since no information will be gained by further solubility tests.

If the compound is soluble in water, test the pH of the solution with universal indicator paper. An organic acid that is soluble in water will give a solution of pH 2 to 3 (indicator paper red). An amine that is water soluble will give a pH of 10 to 11 (indicator paper blue).

If the sample is not soluble in water, repeat the test using 0.5 ml of 5% hydrochloric acid. If the compound dissolves or is significantly more soluble in acid than in water, it can be concluded that it is an amine.

If the compound is insoluble in water and dilute acid, test for acidic properties by repeating the solubility test with 0.5 ml of 5% $NaHCO_3$ solution. If solubility is not observed, test with 5% NaOH to check for a phenol.

Classification and Chemical Tests

At this point, since the scope of the experiment is restricted, it should be possible to place the unknown in one of the six classes of compounds listed in the tables by reference to the solubility data and infrared spectrum. A few chemical tests that give a clearly visible indication of a reaction are sometimes useful to confirm the presence of functional groups that are indicated by spectra. Three reagents for this purpose are described below. These tests should be applied *only as indicated,* and *not* routinely to all unknowns; spurious and confusing results will be obtained, for example, if any of these reagents is applied to an amine.

2,4-Dinitrophenylhydrazine (DNPH)

This reagent is a solution of 2,4-dinitrophenylhydrazine and sulfuric acid in ethanol, and may be useful in the case of *neutral* compounds that contain a **carbonyl** group to distinguish between **aldehydes** or **ketones** on one hand and esters on the other. Practically all aldehydes and ketones give the hydrazone with this reagent, and the rapid formation of a yellow, orange or red precipitate indicates these functional groups. Esters do not react.

Procedure. Dissolve about 0.2 ml or 0.2 g of the unknown in 1 ml of alcohol in a test tube, and add 2 ml of the dinitrophenylhydrazine reagent. If a hydrazone crystallizes, collect it on a suction filter, wash with 1 ml of ethanol and allow the crystals to dry.

Ferric Chloride

Most **phenols** (and stable enols) give a red or violet complex with ferric chloride solution. If solubility and infrared data suggest a phenol, this test can be applied for confirmation if desired. The test should be run first with authentic phenol so that you know the appearance of a positive test.

Procedure. Place 1 ml of 1% ferric chloride solution in a test tube and add one drop or one crystal of the unknown.

Iodoform Test

Methyl ketones ($RCOCH_3$ or $ArCOCH_3$) react rapidly with halogen under basic conditions and are converted to the tri-iodo ketone. This product is then immediately attacked by base and the $-CI_3$ group is displaced to give iodoform, which crystallizes as a pale yellow solid with a characteristic odor. Since hydroxyl groups are oxidized to carbonyl under these conditions, compounds containing the $-CHOHCH_3$ group also give iodoform.

$$RCHOHCH_3 \xrightarrow[KI]{I_2} RCOCH_3 \xrightarrow[KI]{3I_2} [RCOCI_3] \xrightarrow{OH^-} RCO_2^- + CHI_3$$

Iodoform

This test can thus provide confirmatory evidence, if needed, for the presence of $-COCH_3$ or $-CHOHCH_3$ groups in neutral compounds. As in any test of this kind, it is essential to run a known compound along with the unknown to obtain a reliable and meaningful result.

Procedure. Dissolve 3 to 4 drops of the unknown in 2 ml of dioxane in a test tube and add 1 ml of 10% aqueous NaOH. Add dropwise, with shaking, I_2—KI solution (side shelf) until the iodine color persists; 2 to 3 ml is normally required. Warm the solution slightly, and if the color fades, add more iodine. Then remove the iodine color by adding a few more drops of NaOH solution and dilute the solution with 5 ml of water. If the test is positive, iodoform precipitates as a dense yellow solid. To confirm that the precipitate is iodoform, it can be collected and the melting point checked; iodoform has a mp range of 119 to 121°.

PRELIMINARY CLASSIFICATION EXPERIMENT

To gain experience in the classification scheme, solubility tests and inspection of spectra should be carried out with several general unknowns. These compounds are labeled A to E; the infrared and nmr spectra of these compounds are given on pages 222 to 225 at the end of this chapter.

1. Solubilities

Obtain samples of the five compounds in labeled test tubes and go through the solubility tests on each as described; record the data on the report sheet.

2. Infrared Spectra

Examine the infrared spectra of the compounds (pp. 222 and 224); note peaks in the 4000 to 1600 cm^{-1} (2.5 to 6.25 μ) region due to functional groups [4000 to 3000 cm^{-1} (2.5 to 3.3 μ): NH and OH; 1750 to 1650 cm^{-1} (5.70 to 6.05 μ): C=O groups]. Note any confirmatory data from the nmr spectra (blackened peaks due to exchangeable protons: OH or NH).

3. Chemical Tests

Carry out any chemical tests that seem to be required in view of the solubility and spectral data and record the results.

4. Classification

From the combined data classify each of the compounds as (1) aldehyde or ketone, (2) alcohol, (3) phenol, (4) acid, (5) amine or (6) ester. Summarize the observations and conclusions that led you to your assignments.

5. Nmr Data

Examine the nmr spectra (pp. 223 and 225); see if you can deduce the structures of any of the compounds. In looking at the spectra, observe the following points and record them in a systematic way:

a. List each distinct peak or multiplet and the relative number of protons from the integration line. To arrive at the relative areas of the peaks, measure the height in cm^{-1} of each major step in the integration curve and derive the smallest set of integers that are in the same ratio. If there is a clearly recognizable CH$_3$ peak, set the value of this at 3 and derive the closest whole number values for the other peaks.

To illustrate, the spectrum of compound A, for example, contains one group in the 6.5 to 7.0 ppm region and a singlet at 3.3 ppm. The relative heights, adjusted to the closest whole number ratio, are 5:2.

b. Note the position or chemical shift of the peaks or groups, and from Table 20.2 (page 204), assign the type of proton — aromatic, CH$_3$, CHO and so forth.

c. Record characteristic splitting or appearance of peaks, such as multiplet, doublet and so on, and their significance. For example, a singlet with 3 protons in the region 2 to 3 ppm indicates a CH_3 group on a carbon atom with no protons attached.

INDIVIDUAL UNKNOWNS

After completing the preliminary work on the general unknowns, obtain samples of unknowns and spectra from your instructor and proceed as follows:

1. Determine the melting point, if a solid, or the boiling point, if a liquid, unless instructed to omit boiling point.

2. Carry out solubility tests and spectra examination, and classify the compound as to type of functional group.

3. At this stage, review the data, draw all conclusions possible and then examine the appropriate table of compounds. Write out the structures of the compounds in the list that appear to be possibilities.

4. Prepare a derivative to confirm the identification, compare the melting point with that given in the table and submit it to your instructor with your report.

Derivatives

A derivative is a reaction product that is characteristic for the unknown, and provides the final evidence in the identification. The melting point of the derivative is an additional physical constant for comparison, and the fact that the unknown reacts with a given reagent in the predicted way is confirmatory evidence for the functional group.

The derivatives listed in the tables are known, with a few exceptions, for all of the compounds included in each table, and they can be obtained from any of the compounds by simple, standard methods. The derivative should be chosen to distinguish between possible candidates; in a few cases, preparation of two derivatives may be desirable to eliminate any uncertainty. When possible, select a derivative that has a melting point between 100 and 200°. General procedures are given below for each type of derivative, but the final crystallization details will vary somewhat, depending on the solubility of the specific derivative. Recall the generalization that the higher the melting point, the lower the solubility.

TWO EXAMPLES OF IDENTIFICATIONS

To illustrate the overall identification process, two examples are outlined below, with the data as they would be obtained and the conclusions as they would be drawn.

Wave number, cm⁻¹

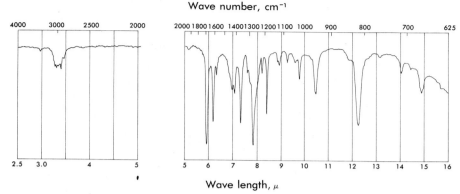

Wave length, μ

FIGURE 21.3 Infrared spectrum.

Example 1. The unknown is a high-boiling oil; the infrared and nmr spectra are given in Figures 21.3 and 21.4.

a. The compound is not soluble in water, dilute acid or dilute NaOH. *Conclusion:* Compound is not an acid, phenol or amine.

b. The infrared spectrum (Fig. 21.3) shows no band for OH, but it does contain a peak at 1690 cm⁻¹ (5.9 μ), indicating a C=O group, probably conjugated with an aromatic system. *Conclusion:* Compound is not an alcohol, but could be an aldehyde, ketone or ester.

c. Reaction with 2,4-DNPH gives a bright red precipitate. *Conclusion:* Compound is an aldehyde or ketone.

d. Nmr spectrum (Fig. 21.4) contains two singlet peaks of equal area at 2.3 and 2.4 ppm and two doublets of equal height in the 7 to 8 ppm region.

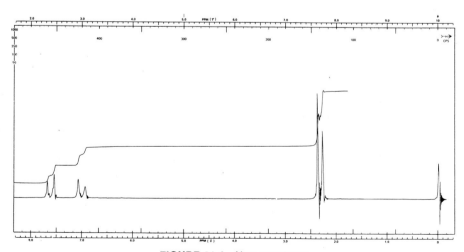

FIGURE 21.4 Nmr spectrum.

The ratio of peak heights from left to right is 2:2:3:3. There is no peak for a —CHO group in the 9 to 10 ppm region, and the compound is therefore not an aldehyde.

The three-proton singlets are in all probability CH_3 signals; the positions are in the region for CH_3CO— or CH_3Ar groups. The two-proton doublets are in the aromatic region (strong bands in the infrared spectrum confirm a benzene ring). The splitting pattern of the aromatic protons suggests a p-disubstituted ring. Putting the pieces together, a benzene ring with CH_3 and $COCH_3$ groups in the 1,4-positions fits all available data; in other words, the compound is probably p-methylacetophenone.

p-Methylacetophenone

e. A further test, the iodoform reaction, could be carried out at this point to confirm the $COCH_3$ group, but this is not essential provided the derivative checks.

f. The semicarbazone is prepared, and the melting point, 202 to 204°, confirms the identification.

Example 2. The unknown is a colorless liquid; the infrared and nmr spectra are given in Figures 21.5 and 21.6. The boiling point range is found to be 113 to 115°.

a. The compound is not significantly soluble in water or in dilute acid or base. The infrared spectrum contains a strong band at about 3300 cm^{-1}, and no bands in the —C=O region. *Conclusion:* The compound is not acidic or basic, and contains an OH group but not a C=O group, so it must be an alcohol.

b. The nmr spectrum contains the following peaks:

Position	Height (mm)	Type	Splitting
3.8	2.5	exchangeable	—
3.2–3.5	2.4	alkyl O—C—H	5-line multiplet
0.7–1.7	26	CH_2 and/or CH_3	complex

The closest whole number ratio of the peaks is 2:2:20, or 1:1:10, suggesting a total of 12 protons. The 1:1 ratio of the —OH peak and the peak at 3.2 to 3.5 indicates a secondary alcohol, i.e., R—CH(OH)—R.

c. Examining the table of alcohols, four possible compounds within the general range of the observed boiling point are:

	OH | (CH$_3$)$_2$CHCHCH$_3$	OH | CH$_3$CH$_2$CH$_2$CHCH$_3$	OH | CH$_3$CH$_2$CHCH$_2$CH$_3$
CH$_3$CH$_2$CH$_2$CH$_2$OH			
1-Butanol bp 116°	3-Methyl-2-butanol bp 113°	2-Pentanol bp 119°	3-Pentanol bp 116°

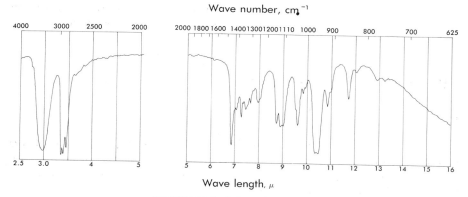

Wave number, cm^{-1}

Wave length, μ

FIGURE 21.5 Infrared spectrum.

d. Two of these possibilities contain the R—CHOHCH$_3$ structure and should give a positive iodoform test. This test was carried out with the unknown, using 2-propanol as a standard for comparison. A trace of precipitate was seen with the unknown, compared to a very strong positive test with the standard. *Conclusion:* The unknown does *not* contain a —CHOHCH$_3$ group.

e. Of the remaining possibilities, 1-butanol does not fit the nmr data; there should be a *two*-proton multiplet for —CH$_2$OH in the 3 to 3.5 ppm region of butanol. On the other hand, the nmr spectrum agrees quite well with the structure of 3-pentanol. Closer inspection of the nmr spectrum in Figure 21.6 shows that the multiplet from 0.7 to 1.7 ppm consists of a distorted triplet from 0.7 to 1.1, with approximate area 6, and a distorted quartet from 1.2 to 1.7, with approximate area 4, corresponding to the CH$_3$ and CH$_2$— groups, respectively, in 3-pentanol with additional coupling due to the CHOH proton. *Conclusion:* The unknown is probably 3-pentanol.

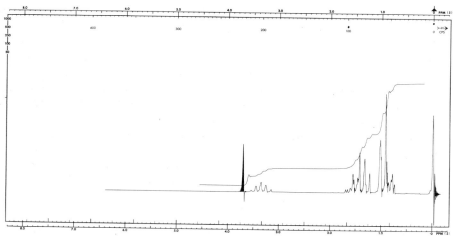

FIGURE 21.6 Nmr spectrum.

f. The 3,5-dinitrobenzoate was prepared, and the melting point range was found to be 98 to 101°, confirming the identity as 3-pentanol.

PROCEDURES FOR DERIVATIVES

Acids

The most satisfactory general derivatives of acids are the **amides** or **anilides** (N-phenyl amides). These are prepared in a two-step procedure by first converting the acid to the acid chloride, and then treating the chloride with ammonia or aniline.

$$RCO_2H + SOCl_2 \longrightarrow RCOCl + SO_2 + HCl$$

$$\underset{\substack{\text{Thionyl} \\ \text{chloride}}}{} \qquad \underset{\substack{\text{Acid} \\ \text{chloride}}}{}$$

$$RCOCl + 2NH_3 \longrightarrow RCONH_2 + NH_4Cl$$

Amide

$$RCOCl + 2C_6H_5NH_2 \longrightarrow RCONHC_6H_5 + C_6H_5NH_3Cl$$

Anilide

Acid Chloride. The first procedure (1) is suitable for most acids, and is the more convenient one. A few compounds such as phenylacetic acid, with a reactive —CH_2— group, require the milder conditions of procedure (2).

1. Mix 1 g of the acid in an 18 × 150 mm test tube with 3 ml of thionyl chloride. Set up a microburner *in the hood,* hold the tube in a clamp and gently heat the mixture to boiling over a low flame. Continue heating for at least 10 minutes at a rate such that most of the liquid condenses on the wall of the tube. The acid dissolves as it reacts; with high-melting insoluble acids it may be necessary to add a little more thionyl chloride and heat longer. When all of the acid dissolves, heat a little more strongly to distill off the excess thionyl chloride (bp 79°). After the solution has been concentrated to an oil, tip the tube sideways to permit the heavy vapors to escape.

2. For sensitive acids, mix 1 g of acid and 2 ml of thionyl chloride in a test tube in the hood as in procedure (1), and fit a calcium chloride drying tube by means of a small one-hole rubber stopper to the mouth of the test tube. Set the tube in a beaker in the hood and shake occasionally during the first 2 hours. If HCl evolution and condensation cause the $CaCl_2$ and cotton in the drying tube to become moist, clean out and dry the tube and replace with fresh cotton and $CaCl_2$. Allow the set-up to stand in the hood for several days to a week. Then evaporate any residual thionyl chloride without heating, using the aspirator.

Amide. To prepare the amide, place 10 ml of concentrated aqueous ammonia and an equal volume of ice in a 50 ml beaker. *In the hood,* add the acid chloride, using a transfer pipet and bulb, to the cold ammonia. After the vigorous reaction occurs, collect the precipitated amide on a

suction filter, wash with a little water and allow it to air dry. Recrystallization is usually not necessary; if the melting point does not agree with that
expected for the derivative in question, recrystallization can be carried out
from a small volume of ethanol plus a few drops of water.

Anilide.   Dilute the acid chloride with 5 ml of methylene chloride and
add it to a solution of 2 ml of aniline in 10 ml of methylene chloride. After
mixing, allow the reaction to stand for a few minutes, add 10 ml of water
and transfer the mixture to a separatory funnel. Add more solvent if necessary to dissolve all of the solid. At this point, it is convenient to add sufficient ether (25 to 30 ml) to make the organic layer lighter than water. (Be
sure no flames are near before pouring ether!) Wash the organic phase by
shaking with 10 ml of 1 N HCl, drain off the aqueous layer and, using test
paper, check that it is acidic. If it is not, repeat the HCl wash and check
again. After removing the aqueous acid layer, wash the methylene chloride
solution with 5 ml of 0.5 N NaOH, drain off the alkaline layer and then
wash with water. Transfer the organic layer to an Erlenmeyer flask and add
a teaspoonful of MgSO$_4$. Shake and allow to stand for a few minutes, and
then filter through a wad of cotton into a clean, dry Erlenmeyer flask. Add
a boiling stone and evaporate the solvent on the steam bath. If the anilide
does not crystallize spontaneously during the evaporation, concentrate to
about 5 ml volume, add 10 ml of ether and cool. Collect the product by
suction filtration and determine the melting point.

Alcohols and Phenols

These compounds can be converted to crystalline **3,5-dinitrobenzoate
esters** by reaction with the acid chloride and pyridine. For a few phenols,
only the benzoate ester is recorded; in this case, substitute benzoyl chloride
in the procedure. Another derivative is an ester of a special type known as a
urethane or carbamate, which is formed by treatment with an isocyanate.

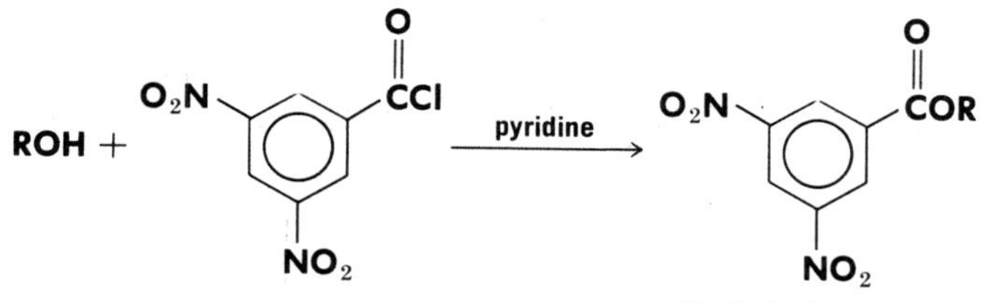

3,5-Dinitrobenzoate ester

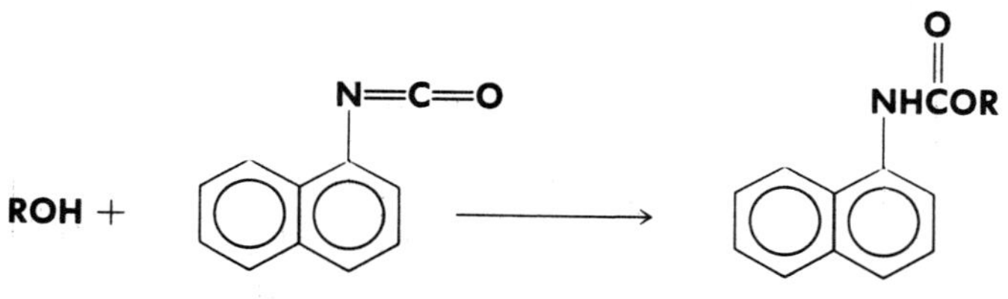

α-Naphthyl
isocyanate

α-Naphthylurethane

Isocyanates are exceptionally reactive compounds, and in any manipulations with isocyanates, the following reactions leading to the diarylurea
will occur if water is present:

$$ArN{=}C{=}O + H_2O \longrightarrow (ArNH\overset{\overset{\displaystyle O}{\|}}{C}OH) \longrightarrow ArNH_2 + CO_2$$

$$ArN{=}C{=}O + ArNH_2 \longrightarrow ArNHCONHAr$$

Diarylurea

The urea is a very high-melting insoluble compound (dinaphthylurea, mp > 250°) and can interfere with the isolation of the desired derivative. Glassware should be dry, and an excess of the isocyanate should be avoided.

3,5-Dinitrobenzoates. Dissolve 0.5 g of the alcohol or phenol in 2 ml of pyridine and add 0.5 g of 3,5-dinitrobenzoyl chloride. Warm the solution gently for 5 minutes and then pour it into 10 ml of water. If a solid ester precipitates, collect this on a filter, wash with water and allow it to dry. Frequently the product separates as an oil. In this case, extract the mixture with ether, wash the ether solution with 1 N HCl as described above for anilides and follow that procedure. The esters generally have rather high solubility in ether; petroleum ether (hexane) is usually a satisfactory solvent for recrystallization.

α-Naphthylurethanes. Place 1 g or 1 ml of the unknown in a *dry* test tube and add 0.5 ml of α-naphthylisocyanate. If the unknown is a phenol, 2 drops of pyridine should be added as a catalyst. Stopper the tube and warm the mixture on the steam bath (do not expose the reaction to moist air!). If the urethane is slow to crystallize, add 2 ml of petroleum ether. If an insoluble white powder is present, this is the diarylurea, and it must be removed by filtering the warm solution. Then cool or concentrate the filtrate to crystallize the derivative.

Aldehydes and Ketones

Semicarbazones and **dinitrophenylhydrazones** are the most common derivatives, and the choice usually depends on the melting point.

$$R\overset{\overset{\displaystyle O}{\|}}{C}H + NH_2NH\overset{\overset{\displaystyle O}{\|}}{C}HH_2 \longrightarrow R{-}\overset{\overset{\displaystyle H}{|}}{C}{=}N{-}NH\overset{\overset{\displaystyle O}{\|}}{C}NH_2$$

Semicarbazide   Semicarbazone

The 2,4-dinitrophenylhydrazone is generally best for aliphatic compounds. For aromatic aldehydes and ketones, the melting point of the 2,4-dinitrophenylhydrazone and sometimes the semicarbazone may be too high for convenience and is actually a decomposition temperature. In this case, the oxime is another possibility. A drawback of oximes is that two stereoisomeric forms with different melting points can be obtained. Melting points of both forms are given in the table; either may be observed.

2,4-Dinitrophenylhydrazones. These derivatives are obtained in the classification test described above. Collect the hydrazone by filtration, wash with ethanol and dry.

Semicarbazones. Mix 0.5 g of semicarbazide hydrochloride and 1 g of sodium acetate in 2 to 3 ml of water, grinding with a stirring rod. When the large crystals have dissolved, add an equal volume of methanol and then 0.5 g of the aldehyde or ketone. Warm the mixture to boiling on a steam bath and then allow it to cool. Add a few drops of water and, if necessary, cool and scratch to crystallize the semicarbazone.

Oximes. Follow the procedure for semicarbazones, using hydroxyl-amine hydrochloride instead of semicarbazide HCl.

$$\underset{\text{Hydroxylamine}}{\overset{\overset{\textstyle O}{\|}}{RCH} + NH_2OH} \longrightarrow \underset{\text{Oxime}}{\overset{\overset{\textstyle N-OH}{\|}}{R-C-H}}$$

Amines

The N-substituted **acetamides** or **benzamides** are easily prepared; the choice will usually depend on which derivative provides the best differentiation between two possible compounds.

$$RNH_2 + \underset{\text{Acetic anhydride}}{(CH_3CO)_2O} \longrightarrow \underset{\text{Acetamide}}{RNHCOCH_3} + CH_3CO_2H$$

$$RNH_2 + \underset{\text{Benzoyl chloride}}{C_6H_5COCl} \longrightarrow \underset{\text{Benzamide}}{RNHCOC_6H_5}$$

Acetamides. Dissolve 0.5 g of the amine in 1 ml of acetic anhydride and heat for a few minutes on the steam bath. Add 1 to 2 ml of water and stir until no oily droplets of acetic anhydride are present. Crystals of the amide usually form on cooling the solution. Collect the crystals on a suction filter, wash with a few milliliters of a 1:1 mixture of alcohol and water and allow to dry. If crystallization does not occur, add ether, transfer to a separatory funnel and proceed as described for the preparation of anilides from acids.

Benzamides. Mix 0.5 g of the amine and 1 ml of pyridine in a test tube, and cool the solution in an ice bath. Add 0.5 ml of benzoyl chloride, stir the mixture and allow it to come to room temperature. Add 1 ml of water and then 1 N hydrochloric acid in small portions until the mixture is acidic to test paper. Stir the solid with a glass rod, grinding up any lumps. Collect the solid on a suction filter and wash first with water and then with 1:1 alcohol-water. Allow solid to dry.

Esters

For the esters that are used as unknowns in this experiment, the most suitable derivative is the **acid** obtained by saponification, although the acid does not provide information about the alcohol part of the ester. Since it

$$\overset{O}{\overset{\|}{RCOR'}} + NaOH \longrightarrow \overset{O}{\overset{\|}{RCO^-Na^+}} + R'OH$$

Sodium salt

$$\overset{O}{\overset{\|}{RCO^-Na^+}} + HCl \longrightarrow RCO_2H$$

is not practical in this experiment to isolate the alcohol and prepare a derivative of it, the alkyl group in the ester must be identified from the nmr spectrum.

Saponification Procedure. Mix 1 g of the ester and 10 ml of 2 N NaOH solution in a 50 ml round-bottom flask. Set-up the flask on a ring and wire gauze and connect a condenser (vertically) for refluxing. Add a boiling stone, start the condenser water, heat the mixture to boiling for 15 to 20 minutes and then cool the flask. If a significant amount of ester (oily

IR SPECTRA OF PRELIMINARY CLASSIFICATION UNKNOWNS

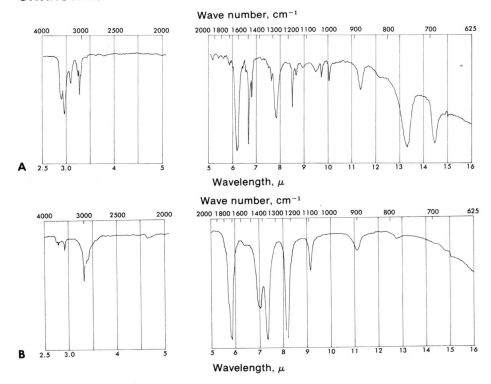

droplets) is still present at this point, add 5 to 6 pellets of NaOH or KOH and a fresh boiling stone and heat the mixture under reflux for another 10 minutes.

With some esters it may be necessary to add 10 ml of ethanol as a solvent. When alcohol is used, it should be removed by distillation before isolating the acid. After refluxing for 20 minutes with aqueous alcoholic hydroxide, allow the solution to cool somewhat and arrange the condenser, with a distillation head and adapter, for distillation. Distill until the volume of distillate is equal to the volume of alcohol added.

Transfer the aqueous solution of the acid salt to a 50 ml Erlenmeyer flask, cool in an ice bath and add, drop by drop, concentrated hydrochloric acid. Stir or swirl the mixture as the acid is added and continue until the solution is acidic to test paper. Collect the acid on a suction filter, wash with water and allow to dry.

NMR SPECTRA OF PRELIMINARY CLASSIFICATION UNKNOWNS

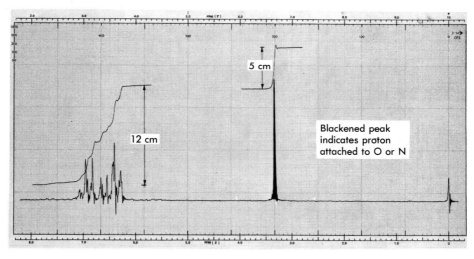

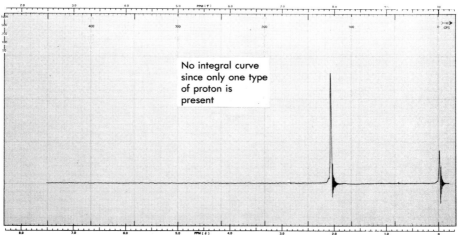

IR SPECTRA OF PRELIMINARY CLASSIFICATION
UNKNOWNS (Continued)

Wave number, cm⁻¹

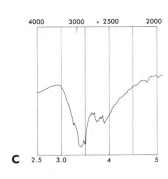
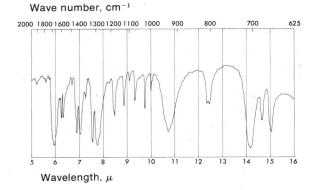

C

Wavelength, μ

Wave number, cm⁻¹

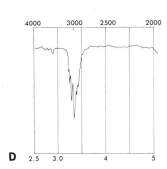
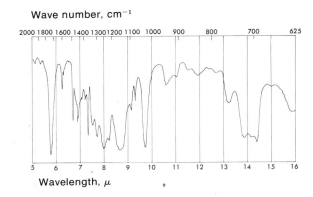

D

Wavelength, μ

Wave number, cm⁻¹

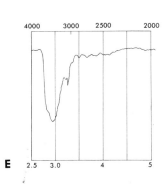
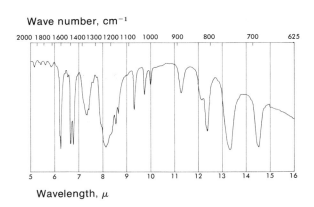

E

Wavelength, μ

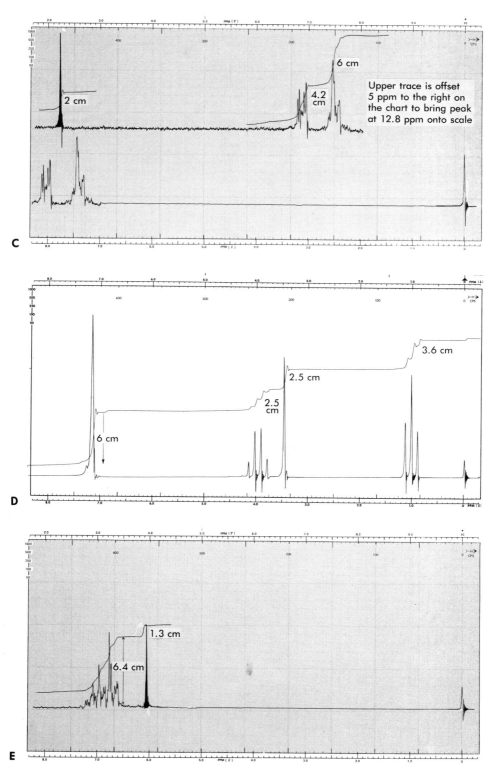

Upper trace is offset 5 ppm to the right on the chart to bring peak at 12.8 ppm onto scale

2 cm

4.2 cm

6 cm

C

6 cm

2.5 cm

2.5 cm

3.6 cm

D

1.3 cm

6.4 cm

E

REPORT

PRELIMINARY CLASSIFICATION EXPERIMENT

1. Solubilities

Mark + for soluble, − for not soluble and 0 for not tested.

Compound	*Water*	*5% HCl*	*5% NaHCO$_3$*	*5% NaOH*
A				
B				
C				
D				
E				

2. Infrared Spectra

Indicate frequency of major peaks.

Compound	*4000–3000 cm^{-1}*	*1750–1650 cm^{-1}*
A		
B		
C		
D		
E		

3. Chemical Tests (2,4-DNPH; FeCl$_3$; I$_2$—KOH)

A

B

C

D

E

4. Classification

A

B

C

D

E

5. Nmr Data

Compound	Position of Peak or Group	Rel. No. of Protons	Chemical Type	Appearance or Splitting	Possible Group
A	6.5–7.0	5	aromatic	multiplet	C_6H_5
	3.3	2	exchangeable	singlet	OH or NH_2
B					
C					
D					
E					

REPORT

INDIVIDUAL UNKNOWN 1

Physical Properties

Solubility Tests

Spectral Data

Infrared

Nmr

Chemical Tests

Conclusions and Possible Compounds

Derivative Preparation

Describe procedure and details of isolation and crystallization, and write complete equation with structural formulas.

Derivative melting point and final conclusions

REPORT

INDIVIDUAL UNKNOWN 2

Physical Properties

Solubility Tests

Spectral Data

Infrared

Nmr

Chemical Tests

Conclusions and Possible Compounds

Derivative Preparation

Describe procedure and details of isolation and crystallization, and write a complete equation with structural formulas.

Derivative Melting Point and Final Conclusions

ALCOHOLS

Compound	Bp, °C (Mp, °C)	Derivatives, mp 3,5-Dinitro benzoate	α-Naphthyl- urethane
Benzyl alcohol	205	112	134
1-Butanol	116	64	71
2-Butanol	99	75	97
Cinnamyl alcohol ($C_6H_5CH{=}CHCH_2OH$)	257 (33)	121	114
2-Chloroethanol	129	92	101
Cyclohexanol	161	112	129
Diphenylmethanol	— (69)	141	136
Ethanol	78	93	79
2-Ethyl-l-butanol	149	52	—
1-Hexanol	156	58	59
4-Methoxybenzyl alcohol	260 (25)		(phenyl- urethane, 92)
4-Methylbenzyl alcohol	— (60)	118	(phenyl- urethane, 79)
2-Methyl-l-butanol	129	70	82
3-Methyl-l-butanol	132	61	68
3-Methyl-2-butanol	113	76	110
2-Methyl-l-pentanol	148	51	75
3-Methyl-2-pentanol	134	43	72
4-Methyl-2-pentanol	132	65	88
2-Methyl-l-propanol	108	86	104
2-Octanol	179	32	63
1-Pentanol	138	46	68
2-Pentanol	119	61	76
3-Pentanol	116	101	95
1-Phenylethanol	203	95	106
2-Phenylethanol	219	108	119
1-Propanol	97	74	80
2-Propanol	83	122	106

PHENOLS

Compound	Bp, °C (Mp, °C)	Benzoate ester	3,5-Dinitro-benzoate	α-Naphthyl-urethane
			Derivatives, mp °C	
4-Chloro-3,5-dimethylphenol	— (115)	(Acetate, 48°)		
o-Chlorophenol	176 (30)	—	143	120
p-Chlorophenol	— (43)	88	186	166
o-Cresol (methylphenol)	190	—	138	142
m-Cresol	202	55	165	128
p-Cresol	— (36)	70	189	146
2,4-Dichlorophenol	— (45)	97	142	—
2,4-Dimethylphenol	— (28)	38	165	135
2,5-Dimethylphenol	— (75)	61	137	172
2,6-Dimethylphenol	— (49)	—	159	177
3,4-Dimethylphenol	— (62)	59	182	141
4-Ethylphenol	— (47)	60	132	128
2-Isopropyl-5-methylphenol (Thymol)	— (51)	33	103	160
2-Isopropylphenol	212	(Aryloxyacetic acid, 133)		
4-Isopropylphenol	— (61)	71	—	—
2-Methoxyphenol	205 (30)	58	141	118
4-Methoxyphenol	— (56)	87	—	—

ALDEHYDES AND KETONES

Compound	Bp, °C (Mp, °C)	Semi-carbazone	2,4-Dinitro-phenyl-hydrazone	Oxime
			Derivatives, mp, °C	
Acetone	56	187	126	—
2-Acetonaphthone	— (54)	234	262	146
(2-$C_{10}H_7COCH_3$)				
Acetophenone	200	198	250	—
($C_6H_5COCH_3$)				
o-Anisaldehyde	— (38)	215	254	92
(2-methoxybenzaldehyde)				
p-Anisaldehyde	247	210	254	65; 133
Benzaldehyde	180	222	237	—
Benzylacetone	235	142	—	87
(4-phenyl-2-butanone)				
2-Butanone	80	146	117	—
n-Butyrophenone	230	187	190	—
($C_6H_5COCH_2CH_2CH_3$)				
Cinnamaldehyde	252	215	255	64; 138
($C_6H_5CH{=}CHCHO$)				
p-Chloroacetophenone	232	201	231	95
p-Chloropropiophenone	— (36)	176	—	62
o-Chlorobenzaldehyde	208	225	207	75; 101
Cyclohexanone	155	166	162	—
2,5-Dimethoxybenzaldehyde	— (50)	208	216*	
3,4-Dimethoxybenzaldehyde	— (44)	177	263	95
2,4-Dimethyl-3-pentanone	125	160	95	—
o-Ethoxybenzaldehyde	248	219	—	—
2-Ethylbutanal	116	99	95	—
2-Heptanone	151	127	89	—
4-Heptanone	145	133	75	—
Isobutyrophenone	222	181	163	—
[$C_6H_5COCH(CH_3)_2$]				
p-Methoxyacetophenone	— (38)	197	231	87
p-Methoxypropiophenone	— (28)	169	—	—
p-Methylacetophenone	226	204	260	88
3-Methyl-2-butanone	94	113	117	—
4-Methyl-2-pentanone	119	135	95	—
1-Naphthaldehyde	292 (34)	221	—	98
2-Pentanone	102	112	144	—
3-Pentanone	102	139	156	—
Phenylacetaldehyde	194	156	121	—
Phenylacetone	216	198	156	—
Propiophenone	218	174	191	—
($C_6H_5COCH_2CH_3$)				
p-Tolualdehyde	204	221	239	80; 110
(p-methylbenzaldehyde)				

*p-Nitrophenylhydrazone

ACIDS

Acid	Mp, °C	Derivatives, mp, °C	
		Amide	Anilide
o-Anisic	100	128	131
(o-methoxybenzoic)			
p-Anisic	183	167	170
o-Chlorobenzoic	140	139	114
m-Chlorobenzoic	156	134	122
p-Chlorophenoxyacetic	156	133	125
2,4-Dichlorobenzoic	158		
3,4-Dichlorobenzoic	208	133	—
3,4-Dimethoxybenzoic	182	164	154
Diphenylacetic	148	168	180
p-Ethoxybenzoic	198	202	169
p-Methoxyphenylacetic	85	189	—
Phenylacetic	76	154	117
2-Phenylbutyric	42	86	—
3-Phenylpropionic	48	82	92
5-Phenylpentanoic	60	109	90
o-Toluic	102	142	125
(o-Methylbenzoic)			
m-Toluic	110	97	125
p-Toluic	117	158	140
3,4,5-Trimethoxybenzoic	170	176	—

AMINES

Compound	Bp, °C (Mp, °C)	Derivatives, mp, °C	
		Acetamide	Benzamide
Aniline	183	114	160
Benzyl amine	184	60	105
($C_6H_5CH_2NH_2$)			
N-Benzylaniline	— (37)	58	107
($C_6H_5CH_2NHC_6H_5$)			
p-Bromoaniline	— (66)	167	204
o-Chloroaniline	207	87	99
m-Chloroaniline	230	72	120
p-Chloroaniline	— (70)	179	192
2,4-Dichloroaniline	— (63)	145	117
p-Ethoxyaniline	250	137	173
N-Ethylaniline	205	54	60
o-Ethylaniline	216	111	147
o-Methoxyaniline	225	87	84
p-Methoxyaniline	— (58)	128	155
4-Methoxy-2-methylaniline	— (30)	134	—
2-Methoxy-5-methylaniline	— (50)	110	—
N-Methylaniline	196	102	63
o-Toluidine	199	112	143
(o-methylaniline)			
m-Toluidine	203	65	125
p-Toluidine	— (45)	153	158

ESTERS

Compound	Bp, °C (Mp, °C)	Derivative, mp, °C Carboxylic Acid
Diethyl phthalate	296	230
Diethyl succinate	216	190
Dimethyl suberate	268	141
Ethyl p-anisate (p-methoxybenzoate)	270	184
Ethyl benzoate	213	121
Ethyl o-chlorobenzoate	255	140
Ethyl cinnamate ($C_6H_5CH{=}CHCO_2Et$)	271	133
Ethyl p-hydroxybenzoate	— (116)	213
Ethyl p-nitrobenzoate	— (57)	241
Ethyl salicylate (o-hydroxybenzoate)	234	157
Ethyl p-toluate (p-methylbenzoate)	241	177
Isopropyl salicylate	255	157
Methyl p-anisate	— (49)	184
Methyl benzoate	198	121
Methyl o-chlorobenzoate	230	140
Methyl cinnamate	— (35)	133
Methyl salicylate	224	157
Methyl p-hydroxybenzoate	— (130)	213
Methyl p-nitrobenzoate	— (95)	241
Methyl phenylacetate	218	76
Methyl o-toluate	213	102
Methyl p-toluate	— (30)	177